Strategy + Teamwork = Great Products

Management Techniques for Manufacturing Companies

Frederick Parker

CRC Press
Taylor & Francis Group
Boca Raton   London   New York

CRC Press is an imprint of the
Taylor & Francis Group, an **informa** business

A PRODUCTIVITY PRESS BOOK

CRC Press
Taylor & Francis Group
6000 Broken Sound Parkway NW, Suite 300
Boca Raton, FL 33487-2742

© 2015 by Taylor & Francis Group, LLC
CRC Press is an imprint of Taylor & Francis Group, an Informa business

No claim to original U.S. Government works

Printed on acid-free paper
Version Date: 20140604

International Standard Book Number-13: 978-1-4822-6010-6 (Hardback)

This book contains information obtained from authentic and highly regarded sources. Reasonable efforts have been made to publish reliable data and information, but the author and publisher cannot assume responsibility for the validity of all materials or the consequences of their use. The authors and publishers have attempted to trace the copyright holders of all material reproduced in this publication and apologize to copyright holders if permission to publish in this form has not been obtained. If any copyright material has not been acknowledged please write and let us know so we may rectify in any future reprint.

Except as permitted under U.S. Copyright Law, no part of this book may be reprinted, reproduced, transmitted, or utilized in any form by any electronic, mechanical, or other means, now known or hereafter invented, including photocopying, microfilming, and recording, or in any information storage or retrieval system, without written permission from the publishers.

For permission to photocopy or use material electronically from this work, please access www.copyright.com (http://www.copyright.com/) or contact the Copyright Clearance Center, Inc. (CCC), 222 Rosewood Drive, Danvers, MA 01923, 978-750-8400. CCC is a not-for-profit organization that provides licenses and registration for a variety of users. For organizations that have been granted a photocopy license by the CCC, a separate system of payment has been arranged.

Trademark Notice: Product or corporate names may be trademarks or registered trademarks, and are used only for identification and explanation without intent to infringe.

Library of Congress Cataloging-in-Publication Data

Parker, Frederick.
  Strategy + teamwork = great products : management techniques for manufacturing companies / Frederick Parker.
      pages cm
  Includes index.
  ISBN 978-1-4822-6010-6
  1. Manufacturing industries--Management. 2. Production control. 3. Industrial management. I. Title.

HD9720.5.P37 2015
658.4'012--dc23   2014020107

Visit the Taylor & Francis Web site at
http://www.taylorandfrancis.com

and the CRC Press Web site at
http://www.crcpress.com

*I dedicate this book to my wife Mary-Anne
who keeps me happy, healthy, and sane.*

Contents

Preface vii

Chapter 1 Role of the Board of Directors.. 1

Chapter 2 Role of the CEO.. 7
　　　　　　Art of Promotion 9
　　　　　　Art of Demotion or Firing...9
　　　　　　How to Survive as a CEO 10
　　　　　　Ten Commandments for the CEO 12

Chapter 3 Creating the Business Plan 13
　　　　　　Strategic Plan..13
　　　　　　Importance of the Right Strategy......................................16
　　　　　　Creating the Business Plan..18

Chapter 4 Corporate Culture.. 21
　　　　　　Deliver on Your Promises 22
　　　　　　Discipline in the Organization..23
　　　　　　Close Supervision 24
　　　　　　Chain of Command 25
　　　　　　Responsiveness..27
　　　　　　Attainable Goals 30
　　　　　　Mission Statement 31
　　　　　　Span of Control...32
　　　　　　Flat Organization..32
　　　　　　Flexibility...33
　　　　　　Summary of Corporate Culture...37

Chapter 5 Communications.. 39

Chapter 6 Organization of Large Corporations 43

Chapter 7 Manufacturing Strategies 49
                Outsourcing and Make or Buy Strategies 49
                Capital Expenditures 51
                Manufacturing Constraints...52
                Cost Reduction 54
                Materials Systems and Supply Chain Management 56
                Office Layouts...59

Chapter 8 Manufacturing 61

Chapter 9 Quality Control 69

Chapter 10 Controlling Overhead 75

Chapter 11 Wage and Salary Administration 81

Chapter 12 Performance Reviews... 89
                Performance Review Form...93

Chapter 13 Motivation and Productivity.. 95

Chapter 14 Benefits and Incentives... 103

Chapter 15 Role of Engineering... 107

Chapter 16 Marketing and Sales.. 109

Chapter 17 Financial Controls... 115

Chapter 18 Human Resources and Training 121

Chapter 19 Checklist 125

Index 129

About the Author... 135

Preface

Manufacturing excellence is not centered only on the shop floor. It is more important to develop the right strategy and foster teamwork between manufacturing, engineering, and marketing in order to be competitive.

Managing a manufacturing company in the global economy is more complex than ever before. In order to be competitive today, the company not only needs to excel in production, but also in marketing, engineering, and every aspect of customer service. Long gone are the days when a manufacturing company could coast along on its manufacturing excellence or its monopoly. When Henry Ford said that the customer could have any color car he wanted as long as it was black, he believed that his company could succeed in manufacturing efficiencies alone. When AT&T had a monopoly on telephones, they could coast along by introducing new products at a slow pace and not care much about what customers wanted. Those days are gone! In today's global economy, the most efficient production in a manufacturing company will fail if it cannot compete in products, quality, costs, and service.

Manufacturing management is about running a business that manufactures, rather than a manufacturing factory. That is why this book starts out at the top, by outlining the duties of the Board of Directors, rather than concerning itself with the factory floor. In my manufacturing career, I became convinced that the "big bucks" were in the right manufacturing strategies and in engineering and marketing, rather than improving the productivity of the workers. That is not to say that the factory efficiency, productivity, and quality should be ignored, but rather that it is not enough to concentrate on only the factory floor in order to become a world-class manufacturer.

This book will show how to best organize a manufacturing company for success. We will get into the nitty-gritty of the manufacturing processes and practices, but only where it has broad application across the majority of manufacturers.

I adopted and embellished a phrase from IBM on the philosophy of management:

> Think big
> Spend small
> Eliminate bureaucracy

Make that your motto also, and you will be successful. In the following pages, you will learn more about how to implement these principles.

The term "supervisor" is used in the generic sense of direct reporting, whether the supervisor is the CEO supervising general managers or a vice president supervising directors or a foreperson supervising lead operators. Thus, the term "supervisor" refers to the relationship of direct reports in the chain of command in a line organization.

1
Role of the Board of Directors

Major duties and responsibilities for the Board of Directors are:

- Serves as trustees for all stockholders
- Develops direction of the company
- Oversees corporate performance
- Selects a CEO
- Approves the Business Plan, including the budget
- Approves any financial proposals that materially affect the business
- Sets up and oversees at least three committees—Audit, Compensation, and Governance

Here are the three committees that are required for the smooth operation and decision making of the Board.

1. Audit committee—Oversees all financial reporting, is responsible for risk and liabilities management, and audits the books.
2. Compensation committee—Sets the compensation for the CEO and officers of the company, including salaries, bonuses, stock options, and any other compensation for key employees. The compensation should reward long-term success and performance, with a sprinkling of bonuses for short-term successes, but very seldom should it be tied to performance of short-term movement of the stock.
3. Governance committee—Draws up and enforces corporate ethical standards, outlines directors' qualifications, outlines their responsibilities, and evaluates the independence and performance of each director. In case of vacancies on the Board, it identifies candidates for the Board.

The CEO should seldom have the dual function of also being Chairman of the Board. The Board must be made up of people who either represent a large investor in the company or can help in some way by using their business connections, skills, or experience in making important decisions for the company. There should be a mechanism for replacing board members when they are no longer useful in these areas. There should be no conflict of interest between board members and their other activities, or friendship with the CEO. (That is easier said than done.) Board members should act in the interest of the shareholders and not in the interest of people in the corporation.

It is not healthy to appoint people to the Board who have so many other responsibilities that they do not have enough time to devote to this particular Board and only serve as a figurehead. Directors should be able to devote sufficient time to carry out duties of the Board. Most importantly, directors should be independent thinkers with no conflict of interest.

It is desirable to have directors serve long terms on the Board because it takes time to become familiar with the business and to have long enough tenure to develop wisdom about the company. However, there should be some limits set up to make sure that directors do not outlive their usefulness. Each company is different, but one possible way to ensure that board members are carrying out their duties is to have age limits (say, 80 years) and to ask directors to resign if they attend less than 75% of board meetings during two years. I specify two years because some directors may have a difficult year and it would not be right to dismiss them if they can usefully serve the rest of the time. As to term limits for directors, I do not believe in that; as long as they perform their duties well, they should remain on the Board.

The Board is responsible for selecting a CEO and for holding him/her responsible for running the company, supplying the vision for the business plan, and being the driving force to implement that plan.

The company must have a strategy outlined in a business plan. The Board cannot do this because not all board members are familiar with the day-to-day operations or the intricacies of the industry. They must delegate that to the CEO. It is the duty of the CEO to have a plan, which he communicates to the Board, and after approval of the plan to communicate that to other employees. The role of the CEO is very important because while he/she will seek input from others, the CEO alone has the responsibility to formulate the plan and execute it.

There are many factors that affect the business—market forces, technology, products, etc. The list is endless. The CEO is the most important factor that can make the difference between success, mediocrity, or failure. If the CEO makes the right decisions, then he/she can overcome the many factors listed above because he/she is better than the average CEO at competitor companies. Teamwork is necessary in a company, but when it comes to the CEO's job, there is no teamwork there. The CEO alone is responsible to the Board. He/she provides the vision for the business plan and is in charge of implementing it.

If the CEO is so important, how much is he/she worth? How should the CEO be compensated?

There has been a lot of press lately about CEOs being over-compensated. It must be recognized that the CEO is different from other managers in the company. All other managers are asked to follow a set of plans, but the CEO must devise the plan. It has been proven that the CEO can make a huge difference in the performance of the company. His/her compensation must be tailored to the size of the company, the type of company, and the difference that a CEO can make to the bottom line of the company.

I don't think benchmarking a CEO compensation package is very useful. It may be better to look for the best person that fits the requirements of the company, and depending upon his/her qualifications see if the Board can afford to hire him/her. During this process, there may be several candidates with different sets of skills and different requirements. There is no set formula for hiring or promoting a CEO. Ideally, there is someone inside the company who can be promoted.

An important consideration is to attract a "long distance runner" rather than a "sprinter." The compensation package should reward long-term performance and not be dependent on short-term stock movements.

There can be special one-time bonuses awarded for reaching some short-term goals, but the overall package should reward performance for the longer term.

While the benefit package can be very generous, no CEO should be hired with a "golden parachute" or exit strategy. If the CEO does not have enough confidence in his/her ability to add value to the business and to be successful in the long run, then the Board should not want that person.

Whether you select a CEO from inside the company or from the outside, your new CEO should hit the road running. He/she should have a track record of upward mobility and demonstrated ability to successfully run a manufacturing company. When you give the specifications to a recruiter or a selection committee, do not fall into the trap of a long list of criteria for Superman. List only four or five of the most important factors. If you select a smart person who knows the industry and has what it takes, you don't need Superman. If the person is smart, he/she will be able to judge the situation and will be able to do the right thing. Do not try to select a CEO for a certain existing situation. Your company at any given time may need a fast turn-around, fiscal discipline, more investments in gaining market share, more new products, or any one of a dozen strategies at the time of your CEO selection. What you are looking for in a CEO is knowledge of the industry and smarts, not special situation skill sets. If he/she is the right person, he/she will listen to the short-term problems and the direction the Board wants to take the company, assess the situation, and make the right choices.

To summarize, your selection criteria should focus on the most important characteristics only.

That is tricky because there have been many studies made as to what are the habits of successful CEOs and the result was inconclusive. Some are quiet, some are shouters, and there are several who have opposite traits yet become very successful, so don't look there for answers.

Most companies choose a marketing executive as the CEO. The reason is that usually they have the best communications skills and as we shall see, communication skills are the most important qualities for a CEO. Marketing executives should have the vision required to create a successful business plan and the necessary communications skills to be successful CEOs. Nonetheless, a manufacturing or engineering executive may be the ideal candidate if he/she possesses the previously mentioned skills. A broad-minded manufacturing executive can make a very good CEO if he/she has a right-hand VP Marketing person to supplement his/her shortcomings in the marketing area. Beware of successful sales executives because often they are better suited for short-term goals rather than longer-term strategies, but this is a generalization and as such does not apply to everyone.

No matter what discipline the CEO comes from, he/she cannot have all the skills necessary to run the business. Don't specify every little skill set needed to run the business and ask for it. The weaknesses can be supplemented by the strength of the CEO's direct reports. The most important strengths needed for the CEO are **vision, knowledge of the marketplace, leadership ability, and communications skills.**

The Board of Directors should set the goals, and a newly hired CEO must agree with them. The goals should be realistic enough so that the CEO—if he/she is that good—can meet or exceed them. No compensation package should give bonuses unless goals are exceeded. That is why goals should be realistic because then there is great leverage in exceeding goals. If goals are set too high and during the time frame it becomes obvious that they cannot be met, there is no incentive for management to work harder as all is lost anyhow.

A word of caution about succession plans. Unless the CEO is getting close to retirement, it is counterproductive to have a succession plan for him/her. A succession plan means an "understudy," someone who is groomed to take the place of the CEO. I just described how broad the responsibilities of the CEO are, and the fact that he/she must be a person of vision. If there is an official "understudy" for this job, it can create a lot of friction within the organization. If the "understudy" is that good that he/she can be thought of as the CEO of the company, he/she can be recruited or can leave if it looks like the current CEO is not close to retirement.

If a "successor" for the CEO were announced, he/she almost certainly would be a member of the CEO's team. That will be the end of the team. Every direct report to the CEO will be afraid to contradict or argue with the "successor" and this can go on for years. If the "successor" is not announced and not told that he/she is the anointed one, it becomes impossible for the CEO to groom the person for the job.

There have been situations where the CEO is close to retirement and some "genius" on the Board decides to set up a few officers within the company to compete for the CEO's job. This is very destructive because it sets up key people and their direct reports against each other, to the point where there are two or three "camps" within the company that try to destroy each other. It is like cancer within the organization. The two or three "successors" will try to undermine each other and the rest of the company will take sides or stand by and watch to see who wins.

2
Role of the CEO

The major difference between managers and the CEO is that the CEO is responsible for the business plan and therefore his/her role requires vision and leadership. While managers work toward goals outlined by guidelines, the CEO sets the guidelines and decides among conflicting strategies what the best course is to follow. He/she then sets these ideas down in the business plan. The direction will sometimes be unpopular and may even conflict with the self-interest of some managers who have to implement them, but that should not stop the CEO to set the right path.

A leader cannot be popular at all times and cannot always please the majority. Great leaders have often broken new ground based on their convictions, rather than listening to consensus. There is an old saying, "If you have your ears to the ground, you cannot stand up and lead." The CEO must assume the leadership role and follow his/her own convictions rather than look for popularity. All successful CEOs have followed this formula. The plodding, reasonable manager, who works on participating principles and consensus, cannot become a successful CEO. Leaders must act on their convictions and be forceful in their methods.

The CEO gives direction. In addition to being a good manager, he/she must also be a competent leader. While the CEO does not take direction, he/she must be open to advice.

Most businesses want to grow sales and profits. That becomes the goal of the CEO. He/she must forge a strategy on how to do that, and then execute this strategy. The CEO's first priority is to create a business plan, get the Board and his/her management team to sign up to that plan, and then execute it. Because every business is different, it is not possible or useful to speculate here what that business plan should look like, but it should show the path to grow sales and profits for the company, and explain why this is the right strategy to do that.

The CEO must create a team to execute the plan. Usually he/she will inherit an executive team of managers. The CEO should limit his/her span of control in order to focus on his/her duties and still spend enough time supervising his/her direct reports, because the CEO must realize that he/she will only be able to execute a plan if his/her direct reports get enough guidance. The CEO must ensure that his/her direct reports, that is, the key management team, are up to the CEO's standards of excellence, and have the competence and drive to make the plan come true. This is no easy task because tough decisions have to be made in marginal situations when a management team is inherited. A good CEO will spend a lot of time training, evaluating, and—if necessary—changing managers directly reporting to him/her.

Span of control of any manager should be limited in order to allow that manager enough time spent with his/her direct reports and to understand what they are doing. This principle applies to the CEO as well. A good span of control limits the number of direct reports to seven and no less than four. That does not include staff, which should not be counted in span of control. If the CEO has too many direct reports, it is time to subdivide the organization into "mini-CEOs" or divisions that have their own organization.

The CEO is the most important person in the organization. After the CEO in importance comes the management team. The level under the management team is next in importance and must be manned by competent managers. I believe that a good CEO must use his/her leverage to influence the selection and retention not only of his/her direct reports, but also those one level below that. For instance, a CEO has a VP Operations reporting to him/her and under the VP Operations is the Materials Manager. The CEO should make sure that the Materials Manager is a good choice and is competent to carry forward the success of the business. The CEO ideally would like to see people two levels down in the organization to be excellent choices in their field of expertise. This is what makes a winning team of people that have the most influence in the success of the business. If the CEO has six direct reports and they all have six lower-level direct reports, the CEO must influence the selection of 42 people, who then constitute the top two tiers of management. This team provides the CEO with the leverage he/she needs to implement his/her vision.

That said, it is often necessary to demote, promote, or hire new people in key positions of the organization. Following are a few techniques of doing this.

ART OF PROMOTION

Promotion is easy and it is fun, but it should come with a price. The promoted employee should be asked to perform at a high level immediately. Objectives reflecting the expectations set for this employee should be outlined at the time of the promotion. The promoted employee should be given full authority showing trust in him/her. This is important because people in the organization will "test the waters" to see how far the authority is given, whether the promotion has the full support of everyone higher in the organization, or whether it is only a stop-gap measure. The right signals must be given to allow the newly promoted employee every chance to succeed. This same principle (of authority given) should also apply to key managers who are new hires.

Sometimes a key position will be vacant and the CEO will allow two or three people to "compete" for that position, while it is filled with someone "acting" in that position. While at times this cannot be avoided, it should not be used as a tactic for motivating people competing for the position. If they are good, they are already working as hard as they can. Competition here does not serve any useful purpose. During this kind of competition, everyone suffers, politics become rampant, and it only serves to the detriment of the organization. Competing for key positions is bad policy.

ART OF DEMOTION OR FIRING

Demotion or firing is the most distasteful part of management. However, if someone does not want to demote or fire people, he/she should never hire or promote anyone. If someone fails in a key position, he/she should be replaced. **No one in a key position should be allowed to fail twice.** Sometimes the incumbent is no longer able to do justice to the position. The choices are to fire, demote, or move the person to another position. Demotion is very seldom a good option because the demoted person will often be bitter and only perform his/her duties grudgingly. Worse than that, he/she may try to undermine the newly selected boss. Demotion may work for someone near retirement who is enlightened enough to dedicate himself of doing a lesser job, but that is very rare.

Moving key people sideways or putting them "on the shelf" (which often happens to useless executives in large organizations) is a bad idea. It sets the wrong example from the CEO, and it creates friction within the organization. The executive who is shoved aside may want to plan his/her "comeback" and undermine decisions at every occasion. At best, it is a waste of money.

Firing a key executive or manager who is no longer able to perform at the highest level should be done with dignity. That means that there should be a mutual understanding on how it will be done, rather than escorting the individual out the door on short notice. Once the decision is made, however, there should not be a long period of lingering death. It is a bad idea to have the fired executive still on the premises when the newly hired person or the promoted individual takes his/her place. When the company decides that an employee is no longer fulfilling his/her function, it is standard practice to "write up" that person a few times and go through the disciplinary measures before firing that individual. This is required to protect the company from lawsuits and possibly to give the employee a chance to fix the problem. This procedure is a lengthy one, and should not be applied to key managers because it is not in the best interest of the company to have someone in a key position being under fire. Too much is at stake! No matter how painful or difficult, once a decision is made to let a key manager go, it should be done swiftly.

HOW TO SURVIVE AS A CEO

The CEO cannot be competent in every aspect of the business. In order to exercise proper management, he/she must recognize his/her limitations, and shore up those areas where he/she is weakest with people who are strong in those disciplines.

The CEO must create an environment conducive to teamwork. This is best done with a line organization with a clear-cut chain of command. There can be no turf fights among key managers. The authority should be with the line organization. Corporate staff should not be competing for authority with line managers. **Corporate staff should mainly be used for auditing duties, not for second guessing line managers.**

The CEO must communicate upward and downward in the organization. Nobody else can do that. He/she must inform and get consent from the Board on every major event or significant changes to the business plan. He/she must communicate the plan and its progress to employees in a way that ensures that employees understand what is going on, and what the company wants from them. The CEO must also communicate what the company is doing for the employees.

The CEO must make sure that he/she and the executives communicate with customers and vendors. Sounds like a lot of communication, but these are essential duties that only the CEO can perform. He/she must decide how much communication is necessary and what to communicate to whom. Tricky business! That is why communication skills are such a key requirement for choosing a CEO.

The CEO leads a lonely life. Everyone in the organization has peers except the CEO. Everyone else has a boss to go to for guidance, talk things over with, or get "chewed out" occasionally, except the CEO. Everyone gets well-defined objectives, whereas the CEO must create and define his/her own objectives. Thus, the CEO must be a self-motivated leader.

President Truman said, "The buck stops here" and that is true of every CEO. What is important is that only the big bucks stop at the CEO, and hundreds of small bucks stop at lower levels. In order to achieve the goal of dealing only with higher-level problems, the CEO must pick executives who are capable of dealing with lower-level problems and will shield the CEO from them, so that he/she can concentrate on the most important issues. Having picked good and competent executives, the CEO then must be able to delegate and not get involved in secondary issues.

The best way for the CEO to handle stress is not to take up yoga, fiddle in an executive sandbox, or embark on the latest stress relief fad, but to delegate with confidence. If the CEO cannot do that, then he/she has the wrong direct reports and must do something about it.

The CEO is hired by the Board of Directors and serves at their pleasure. Therefore, it is important for the CEO to communicate with the Board often and keep the Board informed of any major developments, whether they are good or bad. Nobody likes surprises, least of all the Board of Directors.

Life can be fun at the top, but it requires excellent management and constant vigilance to keep it that way. The following 10 Commandments for the CEO will help to keep the fun going.

TEN COMMANDMENTS FOR THE CEO

1. Hire direct reports to shore up your weaknesses and to manage the most important functions of your organization.
2. Your direct reports should consist mostly of "line people." Use staff sparingly for control and audit only.
3. Limit your span of control to seven. If you cannot do that, it is time to subdivide the organization.
4. Develop a timetable for major functions, like creating the business plan, and make sure that the business plan has attainable goals, that the Board of Directors approve, and your key people buy into it.
5. Keep conflict out of the organization. The best way to do that is to delegate authority where there is responsibility. Insist on following the chain of command, and use it yourself.
6. Spend sufficient time with your direct reports to ensure that their objectives line up with yours and that they are willing and able to execute your plan.
7. Do not hesitate to change your direct reports if they do not live up to your expectations or as business conditions change. **You will only be as good and as effective as the people who report to you**.
8. Ensure that the business is under control and devoid of surprises. Communicate often the state of the business to the Board of Directors.
9. Design incentive systems to keep employees at all levels motivated.
10. Design an MBO program for salaried employees to make sure that everyone is on the same page implementing your plans.

3

Creating the Business Plan

The CEO is responsible for creating the business plan, and getting the Board to approve it. The business plan is like a road map. It defines where the company is now and where it is going within a time frame. There should be a long-range visionary plan spanning 3 to 5 years, called the strategic plan, and a one-year implementation plan that is much more detailed, called the business plan. In case of large corporations with several business units, there should be separate plans for each business unit.

STRATEGIC PLAN

The strategic plan should address the following trends for the next 3 to 5 years:

- What changes will occur in the next few years in your customer base?
- Strengths and weaknesses of your competitors.
- Future technologies in your field, and how that affects your products.
- Future technology changes in manufacturing affecting your production.
- Market channels and market shares for all your product lines.

The strategic plan should be very simple and contain only a few ideas on how to change the present status quo. There is no need to emphasize details. The details belong in the short-range business plan. There is no need to crunch numbers in the strategic plan because that tends to overwhelm strategic thinking.

The strategic plan should review whether a company should be in the business that it is currently in, and how its strengths and weaknesses stack up against competitors. It should address which products the company should invest in, which products it should eliminate, and whether it is manufacturing the right products in the right places competitively.

Trade barriers have fallen and we find ourselves in a global economy, which affects all businesses large or small. This is especially true of manufacturing businesses. If you are manufacturing today, you must be as good or better in every respect as your global competitors. This is a tall order!

The global economy has changed the environment and today a manufacturing business may find that it has the wrong products, may be in the wrong location, or may not have the economy of scale to compete and make a profit. On the other hand, it may have opportunities that open up new markets, or allow it take advantage of new technologies to improve productivity and increase profits. These global views must be incorporated into the vision that drives the strategic plan.

The strategic plan must evaluate how every aspect of your business measures up against the competition, and take corrective action wherever it does not do that. Sometimes it is not possible to compete, and in that case, you must rethink and restructure your business.

Here are some of the factors that should be benchmarked against your competitors worldwide:

1. **Location:** Are you well situated for labor costs, shipping costs, and serving customers?
2. **Size:** Is there an advantage that you have, or are you at a disadvantage due to economies of scale?
3. **Supply chain:** Can you get your raw materials and parts as economically as your competitors can?
4. **Delivery:** Can you deliver products to your customers as well as your competitors can?
5. **Products:** Are your products competitive technologically and price-wise?
6. **Marketing and sales:** How do customers perceive your products and your customer service versus that of your competitors?

You know your business and your customers better than anyone does. You should realistically list your advantages and disadvantages versus your global competitors, and then draw your conclusion on how to go forward.

No matter how good your people are or how hard they work, unless you can compete in the global economy, you are fighting a losing battle.

The correct strategy may be to go for increased market share at the expense of higher profits, relocate factories, outsource more products, or change the product lines for the longer run success of the enterprise. Few are the companies that can continue with their existing business model for a long time without changing their strategies occasionally.

Based on potential, the strategy for a product may depend on the time in its life cycle, whether to increase investment or milk the product. Based on the competitive position of a product, is it right to increase margins or to phase out the product?

Part of the strategic plan should be diversification by acquisition, joint ventures, licensing, etc. The considerations for these strategies are not covered in this book, other than to acknowledge that these must also be part of the strategic plan.

There may be a different strategy needed for each segment and each product. The strategic plan should address these issues and describe the strategies to be used. The business plan should address the tactical implementation.

Following is a checklist to use for completing the strategic plan for each market segment.

Strengths and weaknesses of a segment versus competition:

- Cost to produce
- Distribution
- Promotion—brand name
- Quality
- Volume
- Efficiency
- Geographical location
- Distribution
- Capital strength
- Technology capabilities
- Service
- Customer satisfaction
- Customer service
- Sales presence
- Customizing products

Evaluating strengths and weaknesses, the strategic plan should address the strategy for each segment and product whether to:

- Invest
- Sustain
- Milk
- Phase out
- Withdraw

For the market leader, the strategy should be to keep a step ahead of the competition and have a broad range of products available to prevent a competitor from entering the market. The market leader must maintain its market share and not allow any competitor easy entry.

The market follower must make up its mind whether it is satisfied with that role. If it is satisfied, it must follow the market leader every step of the way. It must realize that it is at a disadvantage against the market leader and must plan on how to overcome that disadvantage. If it competes against the market leader's weak spots, it can maintain good profit margins. If the market follower is not satisfied with being a follower, it must make substantial investments and take the market leader "head on." It must be prepared for the market leader's reaction and for a fight for market share. The risks are very high.

The competitor with a small market share must settle for a market niche. It must be aware of the limitations of its niche and watch for changing conditions of that niche. It must be very dynamic and flexible and cannot settle for a "me too" approach. It must offer the customer something unique within its niche. That could be outstanding service, closeness to the customer, quality differentiation, etc.

IMPORTANCE OF THE RIGHT STRATEGY

An example of how important manufacturing strategy is occurred during my career when I was Chief Manufacturing Engineer of Plantronics. In 1964, Plantronics was a small company in Santa Cruz, California. The product was a lightweight headset and the patent rights were running out. Our largest customer was AT&T, and they wanted Plantronics to manufacture a product designed by Bell Labs. Bell Labs engineers are not known to design for manufacturability and this design was very difficult to make. In addition, the price demanded would make us unprofitable, and they could source their own design anywhere else to squeeze out the best price. I formed a task force with engineering to determine the manufacturability of the Bell Labs designed headset. We found it to be extremely expensive and awkward to use in the field and decided to no bid the design. It was very difficult not to compete for the business of our largest customer, but we did it because we believed that we would suffer the consequences of manufacturing a bad design that the customers would not want. AT&T later subcontracted the manufacture of their headset to our competitor and the product was a failure. The competitor was left with a large loss and we continued to supply AT&T with our own design that was very profitable.

The lesson here is that product design is most important in deciding on the right strategy.

Later that year, our CEO came to me with a proposal from a company specializing in Mexican labor contracts. Our headset production was very labor intensive and we needed to ramp up production for AT&T and others. The proposal to manufacture in Mexico came just at the right time as our patent was about to expire and we were afraid of competition. I was given the project to pursue this "offshore" strategy and found it to be cost effective. This was well before NAFTA, but the economies worked out well and we embarked on a manufacturing strategy of subcontracting for labor in Tijuana, and later establishing our own company there. The factory in Tijuana grew from 4 operators to 700 employees while I was with the company in the 1970s and 1980s, and in 2013, the factory employed 2200 people.

During its existence, Plantronics has withstood stiff competition from Japan, Germany, and later China, and for over 50 years has been the largest lightweight manufacturer in the world. I am convinced that if we had not moved manufacturing to Mexico before our patent expired, Plantronics would no longer be in business today, no matter how efficient our production would have been in California.

The lesson here is that factory location is another key strategic decision that cannot be ignored and swept under the carpet.

A less pleasant outcome of manufacturing strategy that I witnessed occurred at Northern Telecom (or Nortel as it was later called). I was Corporate Vice President of Manufacturing charged with implementing world-class manufacturing techniques in the corporation's several factories. At that time in the 1980s, Northern Telecom had the most advanced manufacturing plants in the world, manufacturing digital switching equipment for the telephone industry. We employed all the latest techniques at the factory floor, but top management failed to recognize that the market was shifting toward computer-driven products. Instead of converting the factories to digital data centers, top management was caught up in the dot com craze and acquired companies left, right, and center at outrageous costs. These acquired companies had different cultures and their products were short lived. Northern could not integrate these companies into its company culture, and tried to salvage the situation by embracing another fad of subcontracting everything, including all of its core competencies. It sold all of its factories and what remained was an empty shell. The stock went from $80 to $2 in short order. Today the stock is worth less than a penny and Nortel has no operations. At the time when it had several high technology manufacturing companies, it could have switched product lines from telephone switching to data switching and data servers before Cisco became dominant in that field. Nortel had the right manufacturing factories to produce data products efficiently, but did not have the vision to switch its thinking from telephony to data. Thankfully, I was not with the company during the time of its demise, but observed it sadly from my next career move.

The previous examples illustrate how important it is to manufacture the right product and how the best and leanest manufacturing is useless in the face of the wrong product strategy.

CREATING THE BUSINESS PLAN

The strategic plan should contain only a few pages and should be a foreword to the business plan.

After the strategic plan is completed, the annual business plan must be formulated with short-term goals and budgets. The Board of Directors must approve the business plan along with the strategic plan. After approval, the CEO must break it down into details for each function and get the staff to buy into the plan.

To create the business plan, the CEO must have a broad view and understanding of the manufacturing business in which the company is engaged and the trends and factors that will affect the business during the time span of the business plan. The CEO must spend the time to gather input about his/her competitors, certainly talk to the Board to get their input, and get information from his/her staff. The CEO must insist that the staff also spend time in formulating parts of the business plan because they have valuable expertise in their disciplines, like manufacturing, engineering, marketing, etc. The annual business plan needs preparation and therefore needs a timetable with milestones on how to arrive at a final draft, rather than becoming an exercise of slapping numbers together at the last minute without sufficient research.

In order to have enough time to devote to the business plan for it to be meaningful and effective, there should be a chain of events leading up to completion and a timetable for these events to formalize the annual planning process.

Following is a typical chain of events for the annual business plan:

- Complete the strategic plan.
- Outline the implementation of strategies for the next 12 months with options.
- Get feedback from your organization.
- Look at alternatives for manufacturing, marketing, and engineering.
- Agree on best options to implement the strategies.
- Formulate a preliminary budget.
- Look for alternative resource allocations within the budget.
- Agree upon the budget and issue the business plan.

Business plans should not be dictated from the top, without using "attainable goals." They should not sound like "Give me a plan for next year that shows 15% increase in profits and 20% growth" without examining whether this is attainable or whether it is in the best strategic long-term interest of the company. You cannot have high profit and high growth every year. Some years you have to invest in the future and sacrifice in certain areas.

It is difficult to define what "attainable goals" are for a given business for a given year, but it is worth the effort to approximate it as well as possible because otherwise the business will make the wrong decisions and the wrong trade-offs. Top-down directives may force the staff to make up numbers that are not realistic and that may cause long-term damage (not going in the right direction due to trying to meet unrealistic goals). On the other hand, if the goals are too soft and easy, there will not be enough effort to achieve great results. **The business plan should be based on attainable goals, the definition of which is excellent execution within the limits of what is possible.**

There is a tendency for CEOs and staff to outline annual plan numbers like "banana curves" or "hockey sticks." In case you have not heard these expressions, they refer to graphs that start out fairly flat and then drastically rise into the fourth quarter of a given year. The fact that governments always do that with their budgets only shows how ineffective they are in planning. Do not allow that to happen to your business. If sales are planned to increase by 12% for the year, it is reasonable to want to see them increase 3% the first quarter and 3% every quarter after that, rather than flat for the first half of the year and then making up the difference by 4% growth in the third quarter and 8% in the fourth.

Correction to the business plan during the year should be avoided. If there is enough time taken to come up with the plan, there is no need for revisions during a given year. On the other hand, the long-range strategic plan should be revised yearly to take into account changing conditions.

The business plan is much more detailed than the strategic plan and contains a lot of numbers and goals. It must be market-oriented and examine the viability and potential of each market segment served with each product. It should delve into the detail of what is the best strategy going forward for each product.

Resource allocation is necessary because even if you have the right strategies for each product, you may not have enough resources to implement all of these strategies. For example, if the right thing to do would be to invest in 12 different new products in the portfolio, there may not be enough money or resources available to do that, and resource allocation must prioritize them and decide which ones to do and which ones to eliminate. The business plan must make the necessary compromises to fit all strategic plans into the budget.

Budgeting is done bottom up, with each department putting together their requirements.

The CEO must ensure that the components add up to the total of the plan. Each budget and each goal for every function must add up to a total sum, which should not exceed the budgeted amount for the year. If they do not add up to that, adjustments must be made top down until they do.

4
Corporate Culture

While each business is different, there are general management principles for corporate culture that apply to all of them. I will detail some of these here. The list is by no means complete, but it includes most of the common elements that make the difference between failure and success in companies whether they are large or small.

There is a 3000-year-old quote from the book of Ecclesiastes: "There is nothing new under the sun."

Industrial nations have been manufacturing products for several hundred years in different countries, and different cultures. Yet, during my 40 years of tenure in industry and in studying management techniques, I came across a new fad every few years that was supposed to improve the success of manufacturing companies. Management fads are the curse of the business world. As inevitably as night follows day, the next fad will follow the last. I have seen autocratic management styles, permissive ones, teamwork-oriented styles, sensitivity training, psychological testing, zero defects, Deming principles, ISO 2000, Total Quality Management, Six Sigma system, and Lean Manufacturing. More details on these systems will be discussed in Chapter 8.

All of these fads have some value (except ISO 2000), but it is dangerous to think that one slogan or one system can replace a multifaceted corporate culture. These programs deal with efficiency and quality on the factory floor, and they will be described in detail Chapter 8. It is important to note that factory floor efficiency is not where the action is in a manufacturing company. If the company is sailing in the wrong direction or tries to manufacture a product that is too costly to make, then improving factory efficiency is akin to rearranging the deck chairs on the Titanic after it hit the iceberg.

Over my years as a manufacturing executive, I became convinced that it is very important for the Manufacturing Manager or the VP of Operations to persuade the CEO and his/her peers that the biggest impact on product costs and quality is not found on the factory floor, but in the correct strategies of the manufacturing business to come up with the right product that is easily manufactured. Therefore, marketing and engineering plays a larger role in the success of a manufacturing company than the efficiency on the factory floor. It then should become obvious that teamwork and cooperation between top managers of manufacturing, marketing, and engineering are essential. The CEO's job is to have the right strategy on where the product, its subassemblies, and parts are manufactured, but even the CEO's best efforts will fail if the product strategy is wrong or if the product is not designed for low-cost manufacturing and the quality built into the design. After that, it is the job of the Manufacturing Manager to implement these strategies and to have the product manufactured efficiently. Of course, that last part is also very important, but it is the tail end of the chain of events.

Here is a quote from a leadership seminar:

> Companies are like families in the sense that if the parents get along, then it is likely that the rest of the family will be relatively harmonious. But if the parents do not get along, it is highly likely that there is going to be conflict in the rest of the family—that to some degree mirrors the conflict between the parents. If the executive team is talented and unified in their approach, treats each other with respect and communicates openly, their behavior will be mirrored by the rest of the company.

If each department operates in a vacuum without much interface or consultation with others, no matter how talented these executives are the result will be failure to compete.

It is up to the CEO to promote open communication and a teamwork approach to the different disciplines under him/her.

DELIVER ON YOUR PROMISES

The difference between manufacturing and other disciplines in the organization is that manufacturing has to deliver everything that is promised.

Sales may skip a potential customer, engineering can prioritize what they are going to develop, marketing may neglect a product line, but manufacturing cannot pick and choose. Not only do they have to deliver every order, they also have to deliver it at the time that it was promised. Sometimes these promises are made without consulting manufacturing. This then can make the difference between excellent companies and average or below average ones. The really good companies deliver what they promise. This is done by a corporate culture that starts at the top by ensuring that the whole team agrees on what the deliverables are and what can be achieved and completed in a given time frame. A good company does not promise release dates that will not allow engineering to fully test the product. A good company does not promise features that manufacturing cannot incorporate into the product in time.

An example of an excellent new product introduction is the Tesla automobile. In the past, car companies rushed to introduce new models and some of them had shoddy quality and problems initially with many customer complaints. Tesla introduced a brand new concept in a very complex technology and it could have rushed it to market and promise delivery dates like its competitors in the automobile industry used to do, but it chose not to do that. After Tesla introduced its brand new automobile, the company got awards for customer satisfaction during the first year. The company does not promise deliveries it cannot meet and customers love the car and the company.

DISCIPLINE IN THE ORGANIZATION

A business is not a democracy. It is a benign dictatorship. There must be discipline in a manufacturing organization. Management must be tough because in the global economy, competition is tough. Employees in companies with tough management enjoy growth opportunities, while employees in loosely run companies may get lucky for a while, but eventually that luck runs out. However, an organization should not be hard headed. It should not set impossible goals and rule by fear, nor should it have nitpicking management that counts paper clips. The distinction between tough and hard management makes a difference in how well the company is run. Leather is tough. It gives a little but remains flexible. Glass is hard and inflexible, but a good blow will shatter it. Laissez-faire management is like putty; it is soft and deforms at the slightest pressure. Make sure your company culture is like leather—tough but flexible.

I believe this can be achieved if management sets out to have attainable goals and fair compensation, and insists on excellent performance. **A tough company sets standards for everyone, measures people, and expects them to live up to these standards.**

Given fair compensation and attainable goals, most people do not mind rules and will follow their leaders.

Our society is caught up in the dilemma of discipline versus permissiveness. In industry, permissiveness cannot be the mode of operation. A corporation is not the right place for social experiments. Laissez-fare management has never succeeded in the corporate world, and especially not in manufacturing. Manufacturing is hard and requires discipline. "Discipline" is not a bad word and in this connotation, it means insistence on teamwork, responsiveness, data integrity, formalized procedures, close supervision of employees, and respect for the chain of command.

Discipline in the organization means that there are regularly scheduled audits for everything that needs constant vigilance. It means that people are given realistic estimates to complete assignments and are held accountable for it.

CLOSE SUPERVISION

In order to have discipline and productivity in an organization, there must be close supervision of every function and every employee. Everyone needs a boss to tell him/her where he/she stands, to praise his/her accomplishments, to urge him/her to do better. Some people work better under pressure, others need a little nudge, but nobody works at his/her best without some pressure from the top.

By close supervision, I do not mean standing behind an employee. I mean setting standards, measuring output, and giving feedback. I mean leading by example and frequent communications. If the boss comes in late and goes home early, don't expect employees to strive for extra effort during the time he/she is not around. My motto to floor supervisors is always that they have to be "first in, last out."

If supervisors spend a lot of time with their subordinates, it gives them little time to do any work themselves. This is actually desirable because the role of supervisors is to leverage the efforts of the people they supervise. If they work a lot, the leverage will suffer. If floor supervisors are not expected to do much actual work, but concentrate on supervision, it is necessary to have some working leads appointed in order to give supervisors enough time for supervisory duties. A supervisor of seven hourly people does not need eight hours per day to supervise those seven employees. That is where it becomes necessary to have working leads. An example would be assembly lines where a supervisor can supervise six assembly lines, each assembly line having a working lead with seven assemblers, meaning that a supervisor can supervise 42 people.

In hi-tech companies or companies with much profit, there has been a tendency to form work groups that have very little supervision. Some employees are allowed to work from home, some are working part time, and in some cases, they may not be accountable for their output. This may work for a while for companies with much profit, or where creativity is more important than output, but it is rarely the case in a manufacturing company. Manufacturing companies should stick to their knitting.

CHAIN OF COMMAND

In a well-run organization, the corporate culture should strongly support the principle of chain of command. There are numerous stories and anecdotes of individuals not respecting authority and achieving great successes in the corporate world. There are theories that within the corporation a "maverick" spirit should be fostered. The reason for going against the entrenched order can only be justified when the entrenched order is too bureaucratic and there are too many levels for decision-making. It is important to tackle the key problem here, rather than allow the organization to sink down into chaos. If the chain of command is not working or there are too many levels of authority and the organization does not allow dissent, the organization will eventually fail.

There has to be a corporate culture that allows resolution of conflicts within the organization without cutting across lines and jumping over the chain of command. It does not happen by itself. The corporate culture should address the problem of allowing individuals within the organization to express their ideas and to resolve their differences. There are two reasons why chain of command can be obstructive. One is when individuals cannot be heard. The other is when conflict among peers cannot be resolved. This usually happens when different departments are pulling in different directions. The cure is to have a corporate culture that fosters teamwork between departments and allows individuals to have their say without fear of punishment.

It is not enough to express thoughts of cooperation and freedom of expression. The Human Resources training program and the CEO should educate managers and professionals in the organization to adhere to these principles of teamwork. They do not come naturally. They are counterintuitive because human nature will sway department managers to be protective of their turf and their authority.

Why is protection of turf bad? Because in a dynamic organization it is essential to form task forces, which takes the best people away from their departments. A dynamic organization requires matrix management for many disciplines, and this can lead to conflicts of interest. These conflicts have to be resolved; otherwise, they become a cancer. When two peers cannot resolve their conflicts, there should be a way to elevate the conflicts to the next level of supervision until they are resolved, or finally elevated to the CEO. If conflicts are resolved, there is no need to jump over the chain of command.

In a large corporation, conflict can also arise between divisions, especially when there is a transfer of products between them or when they compete for corporate resources. There should be a mechanism for corporate arbitration of these disputes.

When I was General Manager at Northern Telecom switching division, we were told that we should be buying telephones from our telephone division and special semiconductors from our semiconductor division. These components were priced at transfer prices, which we could not agree on. My switching division was asked to continually increase our profits, yet these internal components were priced higher than what we could buy outside the corporation. There was a constant battle going on between our divisions. Sound familiar?

Corporate stayed on the sidelines to let us battle it out, but forbade me from buying from the outside. I was told to get competitive bids and try to force our other divisions to be competitive in the end. We wasted a tremendous amount of management time haggling over details of deals because there was no resolution. The other divisions knew that I could not go outside, argued many reasons why competitive bids were not equivalent, and asked for enough time to get their costs down.

If Corporate had arbitrated the pricing because strategically they wanted the component divisions to have enough volume for survival, it would have saved a lot of time and goodwill. As it turned out, the component divisions were not competitive and eventually had to be shut down. There was no need to subsidize them by our system divisions. A corporate arbitrator could have foreseen this and saved the corporation a lot of money and wasted management times.

RESPONSIVENESS

Responsiveness should be part of the corporate culture in order to serve the customers and to serve "internal customers" within the corporation better. It is the type of behavior that should be expected in a corporate culture that wants to excel.

The following examples will illustrate how destructive lack of responsiveness can be, and how easy it is to cure it.

An office receives an average of 200 inquiries every week. It has a 10-week backlog of 2000 inquiries. It can process 250 inquiries in a productive week, but due to vacations, holidays, and sickness, it only handles 200 on average. When there is a flood of inquiries, the office works overtime in order to keep the backlog at 2000. This office can work for years without reducing its backlog, which then becomes the norm while the customers suffer.

If it would get temporary help plus some overtime, it could reduce its backlog to one week. After that, it should never allow the backlog to build above 200. This way it could reduce its response time by 900%. A backlog of inquiries of one week or less should be the norm in the right corporate culture.

Another example concerns my experience as a consultant. The company had a repair operation for items under warranty that have been returned by customers. The repair operation was small with only one telephone to answer queries and it was overwhelmed by too many people calling.

Customers were getting angry because of lack of responsiveness, and the company did not want to add another person to the unprofitable repair department. We installed automatic answering devices directing customers to frequently asked questions and websites, but in order to take care of overflow inquiries at times the one operator was still not enough. We solved that problem by rerouting the overflow calls to customer service. Customer service had several operators who were not always busy at the same time. After some minor training, they could answer most questions about repair, and the workload evened out without having to add anyone. Most importantly, customers were not left without answers.

Internally the most controversial areas of responsiveness are engineering change requests and IT requests. There never seems to be enough time or labor to satisfy all such requests. My answer to that problem was to have a standing committee to prioritize the never-ending workload and then schedule completions, and notify the "requestor" of the disposition and status of the request. In some cases, the answer was "No we will not do that" and a short reason why was given. This at least stopped the matter from festering in no man's land and it prioritized the requests, rather than having them done in the order of receipt. This is important because the engineering department or the IT department cannot know what the impact is on the company if requests are not promptly acted upon. The standing committee met once a month and all functions were represented to make their case for priorities.

Another internal discipline of serving people inside the company is promising actions or dates to others. The chairperson of a meeting should expect attendees to come prepared to meetings having met the deadlines they promised and being prepared to answer the questions posed at the last meeting. Wouldn't that be a miracle? My experience has been that people assigned to a task force have many responsibilities and often when they come to meetings of the task force to answer questions or meet deadlines agreed upon, they promise to do it later, but in fact have not done their assignment. It is up to the chair of the meeting to try to get results, but he/she cannot do that unless the corporate culture teaches responsiveness and he/she gets support from the department heads to correct the behavior of non-responsive task force members.

A very good example of how valuable responsiveness is to customers is the example set by Amazon. It swept away the competition by its responsiveness to customers and is continuously in the process of improving delivery times.

Corporate culture should include training and explanations of why response is important and how to provide fast responses. Here are some proposed guidelines, but this list is only partial. There are many more examples of required responsiveness.

- Phones should be answered after less than 4 rings (2 is better, but see below).
- Letters or e-mails should be answered within 6 days (3 is better, 1 is best).
- Engineering change requests must have a disposition within 15 days.
- Repair must be returned to customers within 15 days.

When it comes to answering telephones, technology has advanced to the point where this becomes a complex issue. We now have cell phones, e-mail, and websites in addition to landlines. Frequently asked questions can be answered automatically. Some companies cannot possibly afford to have calls answered from every caller. Of course, a manufacturing company cannot be compared with Microsoft in terms of the number of calls that need attention, but they should emulate Amazon rather than Microsoft when it comes to customer service.

Common sense must be used on how to be responsive to customers and to each other within the company, by using innovative technology and issuing guidelines on what responsiveness is expected at various functions.

Many critics will say that in today's complex world you cannot respond to everything. A salesperson cannot call on every customer in his/her territory. The engineering department cannot develop new products and also update and improve all existing products with engineering changes. The human resources department cannot interview all applicants and send them personalized letters of rejection, and run quality circles, suggestion programs, and babysitting service. The critics are missing the point. A good manager should not start things he/she cannot finish. He/she should understand the priorities, and communicate them within the organization to set the right expectations. For instance, if there are too many inquiries about products that telephone operators cannot handle, an automated system with frequently asked question can relieve some of the load. Some companies find it impossible to answer unsolicited telephone calls and must consciously prioritize how and to whom to respond, but in that case, it should be a policy decision and not benign neglect.

Each department has its own specific activities that require response and must devise their own guidelines to fall in line with these principles.

Do not accept excuses like vacations, peak loads, budget constraints, etc. If you do not have enough budget money to be responsive, that backlog will just keep growing and growing until it blows up in your face. Something should be done! If there is not enough money and there is no technical solution, there should be agreement to cut some of the activities that you cannot fulfill properly.

ATTAINABLE GOALS

Throughout the book, the term *attainable goals* is used often. In some organizations, those who promise the most or shout the loudest get the promotions. This is based on the idea that even though goals are impossible to meet, people under pressure will stretch and try harder to meet those goals. I do not believe that this policy leads to long-term success because it does not promote a healthy corporate culture. I believe that it can be counterproductive because as people are halfway through the target and see that the goals cannot be met, they will give up trying.

I urge companies to adopt attainable goals in their culture, starting with the business plan, down to goals for professionals, and a basis for measurements and incentives for hourly employees.

It is important to define attainable goals. "Attainable" in this context means difficult to achieve, but possible to exceed with excellent performance, all else being equal. Naturally, outside influences and luck play roles in achieving any objective.

Attainable goals should become a corporate culture, explained in easy to understand terms and enforced throughout the organization.

When you use attainable goals, it becomes much easier to plan and forecast and that makes the organization more responsive, more profitable, and devoid of surprises. If goals are set too high (or too low), all kinds of inefficiencies will result due to poor planning. An example would be if forecasted sales are higher than those that materialize, the company would get stuck with too much inventory because the materials department will buy according to the plan and it takes time to adjust for reality.

I know of an executive who said, "I give my people impossible goals and when they meet them I raise the bar." His company is no longer in business.

MISSION STATEMENT

One slogan that many corporations like to embrace is a mission statement. Often these statements sound trite and uninspiring, and contain meaningless lofty phrases that appear hypocritical because the organization does not measure up to them. Language such as "providing high level of service to customers" or "striving to increase shareholders' value" sounds impressive, but what does it mean? No corporation sets out to "provide low level of service to our customers" or "striving to decrease shareholders' value." The test is if the words can be used in anybody's mission statement, then it has no meaning or purpose. The mission statement should be specific to the company in order to be meaningful. It should reflect highlights of the business plan. Thus, it should change over time and should not be etched in stone. The mission statement that comes down in a tablet form from the Board or the CEO is not going to work because it does not engage the organization. In a good corporate culture, the business plan is forged from bottom up, approved top down, and the mission statement should reflect the plan. Thus, when the plan is communicated to the organization, the mission statement becomes meaningful and useful.

Here are a few sentences that belong to a meaningful mission statement that is specific to an organization:

- We will rapidly exit the market for products or product lines that become commodities and thus can only compete on price.
- Our brand name implies the highest quality available. Therefore, we will only manufacture clothing using the best available materials and workmanship and will not compromise by introducing lower-grade products even under different brand names.
- We will strive to be market leaders in all of our product lines. Unless our product line is in the top three market leaders or has a good chance to get there, we will exit that market and search for opportunities elsewhere.

One final word—it may be redundant and unnecessary to have a mission statement because the business plan cannot be condensed into a few words. The usefulness may be only to the outside world for public relations (if that is necessary).

SPAN OF CONTROL

If span of control dictates too many levels, that rules out the goal of having a flat organization. In that case, the entity should be broken down into profit centers that are more manageable and will result in both good supervision and a flat organization. These divisions are further explained in Chapter 6, where it is most applicable.

While each type of manufacturing is different, some general guidelines can be used for good practices for span of control. For salaried employees, a good rule of thumb is 4 to 7 people. For the production floor, one supervisor can supervise up to 15 people comfortably. If you have working leads on the production floor, one supervisor can supervise up to 6 working leads and this way have a span of control of up to 50 people.

Sometimes at the top end of the chain, this ratio breaks down. If a department manager has only two or three managers, how can you have a flat organization? The answer is that the department manager should have some lower level supervisors also report to him/her. In a smaller organization, there may be no need for middle managers, only department managers and supervisors.

Span of control guidelines refer only to line organizational people, while staff functions can be additional reports who require little supervision.

FLAT ORGANIZATION

Good corporate practice should strive for a flat organization. This means that there should be as few levels of management as possible. There seems to be a conflict between this objective and keeping the span of control relatively small, for example, 4 to 7 direct reports. It is important not to have too many people reporting to one supervisor because otherwise, there can be no assurance that he/she can effectively supervise them. Supervisors need to use the leverage of their position to get the most productivity out of their people. If there are too many people reporting to them, they cannot be effective.

Of course, every situation is different and when you start to flatten the organization, you will always find that when you arrive at the bottom or top of the ladder there are not enough people left to have the right span of control. The important thing is to keep in mind the trade-off between wide span of control and flat organization and always strive to do both.

Decision-making must be driven to the lowest level qualified to make that decision. Part of the advantages of a flat organization is to ensure that problems and decisions are handled on the spot and resolved quickly, without too many levels needed to make the call. There should be no more than two signatures required to approve most requisitions or to make most decisions.

On the following three pages, there are examples (Figures 4.1 through 4.4) of how to improve an organization at the top and in the middle management levels to have a reasonable span of control and to achieve a flat organization.

FLEXIBILITY

Part of corporate culture should be to create a flexible organization, where people are not restricted in movement across organizational lines. Organizations need task forces and project teams that are constructed from different disciplines within the organization. Management should have the ability to form these task forces and get the best and most qualified people on them, until the task force is dissolved. In a company with a healthy corporate culture, the departmental barriers do not stop the formation of these groups. There is no room for little kingdoms within a corporation.

34 • *Strategy + Teamwork = Great Products*

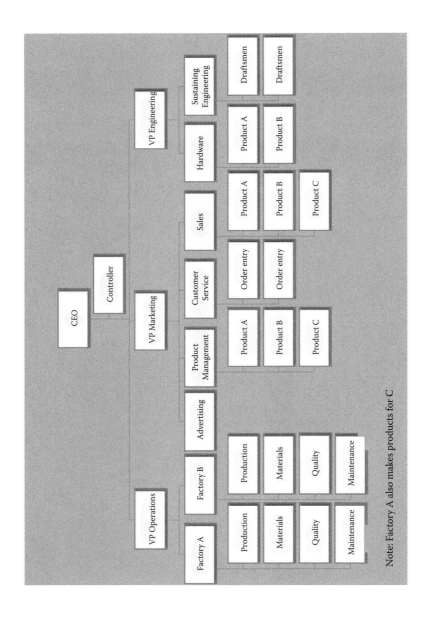

FIGURE 4.1
Top management organization with too many levels.

Corporate Culture • 35

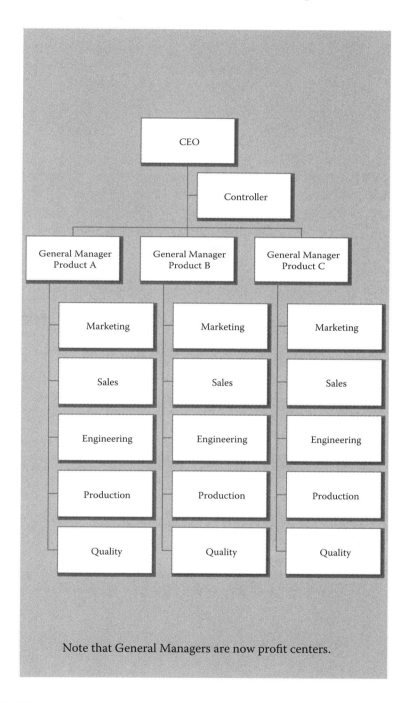

FIGURE 4.2
Top management improved organization resulting in a flatter organization.

36 • *Strategy + Teamwork = Great Products*

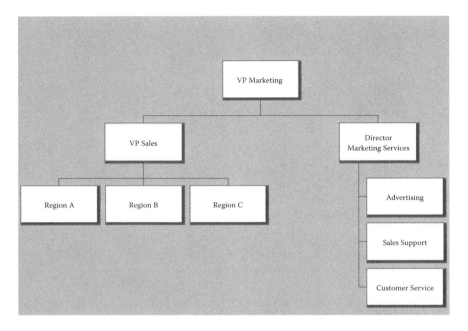

FIGURE 4.3
Middle level organization with too few reporting to VP Marketing.

NOTE THAT THE REORGANIZATION HAS RESULTED IN A FLATTER ORGANIZATION

FIGURE 4.4
Middle level organization improved VP Marketing now has 6 direct reports.

SUMMARY OF CORPORATE CULTURE

To summarize, a desirable corporate culture should set out to implement the following:

- Discipline in the workplace
- Close supervision
- Chain of command principles
- Responsiveness to customers and within the organization
- Attainable goals
- Meaningful mission statement or none at all
- Manageable span of control
- Flat organization
- Flexibility to move across departmental lines

5

Communications

Communications in a corporation is crucial to running a business that is in touch with everybody (or at least most people) in the corporation, and to ensure that they are rowing in the same direction. In order to put some discipline into internal communications, it is important to educate employees on how to communicate with each other inside the company. Employees come from different companies and cultures and it is too much to expect that they all know how they are supposed to communicate inside the corporation.

Education and training of employees is usually a Human Resources job. I suggest that Human Resources should have an educational course repeated every year on communications within the organization including Internet etiquette, conducting meetings, and telephone usage.

The simple principle of e-mail etiquette should be that if you want to get something done, send it to the person who can do something about it. **Do not copy anyone else.** If the request is not successful and you want to reinforce your request, you may send a second one with a copy to the supervisor. If your e-mail is informational in nature, select carefully whose work will be affected and copy only them. Do not create groups of people to automatically send them information regarding a subject or topic. I heard of one junior executive who complained that he came to work some days and spent all morning reading his e-mail in-box and all afternoon answering e-mails.

The biggest problem today is the Internet, which is an open window not only to the inside world but also the outside world. Little can be done to stop employees from surfing the Web, but it should be recognized as the biggest time waster of them all. A lot can be done to stop people from copying each other on internal communications. It is customary in corporations for employees to send e-mail copies to many people to "advertise" their accomplishments, to make sure that no one is hurt by being excluded, or to CYA.

We are all suffering from information overload and corporations are no exception. The corporation must find a strategy to minimize this overload, otherwise the organization will not be efficient. Ask your subordinates not to send you informational e-mail copies unless they are absolutely sure you need to know it. Use the unsubscribe button freely; a little effort up front will save you a lot of time later. When you get e-mail copies that you don't want, send them back to the originator with a polite note asking to be deleted from such notes.

Another means of communication is the telephone. Employees should be trained to answer telephones no later than the third ring. Customer service people should not put their phones automatically on answering machine mode and then gather the messages to answer them at their convenience, unless this is a policy decision.

One way to discipline communications within an organization is to formalize meetings into a schedule of weekly and monthly meetings and issue a matrix of these schedules to avoid conflicts. Someone in the organization should be in charge of setting up meeting schedules. The higher up in the organization the better. It could be the Administrative Assistant to the General Manager. Do not leave it to chance! Also, include in employee education a course on how to conduct meetings and how to come prepared for meetings.

Meetings are a crucial communications tool, but it should be understood that they are a great time waster as well. Many people in corporations complain that they are asked to attend too many meetings and therefore cannot do their own job. That is why it is important to know how to conduct meetings and to keep attendance down to only those who need to be there.

Every meeting should have a chairperson to keep it in context, and an agenda issued to participants before the meeting. Participants should be restricted to only those who can contribute to the meeting. If they only go there to know what decisions will be reached, they can be notified later without wasting their time attending the whole meeting. The chairperson should have some basic training in how to keep the meeting on agenda and how to keep it short.

One of my favorite expressions in communication is, **"Completed staff work."** This concept should be taught throughout the organization and be insisted upon by key managers. Completed staff work reports on a problem possibly show alternative solutions, but always recommend a favored solution or an action to be taken. All a supervisor has to do with completed staff work is to approve or disapprove it and, if necessary, send it on higher in the organization without having to do anything else because the work is complete.

A corporation must find time to listen to its employees. This way tensions can be diffused before they become threats, and employees need to know that someone is taking them seriously by listening to them. Input from employees can be useful in solving problems or saving money. The corporation should devise a "bottom-up" communication program as part of its corporate culture. For example, it should be the practice of every supervisor to have "listening sessions" from small groups of his/her reports. I emphasize small groups because that is the most effective way to allow people to voice their opinions/concerns. In large groups, no one speaks up other than those whom you probably do not want to hear from, the people who always out-shout others. It is the shy ones and those reluctant to speak up that you want to hear from.

Listening sessions should never include top-down communications. The supervisor can hold listening sessions either on-site or at luncheons. These listening sessions should not be restricted to salaried employees. Line supervisors should also be required to hold them on the factory floor with time devoted during regular working hours. The payoff will be worth the time.

I do not advocate having suggestion plan programs with rewards. These have been tried and (in my opinion) have more problems than they are worth. If you have suggestion programs, you are obligated to have someone investigate every suggestion and reply to it. This someone is usually a manufacturing engineer who should have larger projects to work on than those suggested by employees. His/her replies will be criticized by the "author" of the suggestion. Awards will lead to controversies. There will be more dissatisfaction with the scheme than it has value. The claim that it helps to retain employees with pending suggestions is very suspect.

6

Organization of Large Corporations

We will examine the best ways to organize a large corporation in the global economy. Not all of it will apply to medium-sized or small businesses, but they will find that they can learn a lot from these pages on how to organize most effectively.

A large corporation became large because it did some things well; otherwise, it could not have grown. Still, large organizations need to reinvent themselves from time to time and reexamine their organizational structure or else they become too cumbersome or outmoded.

There is consensus that small organizations are more efficient and easier to manage than large ones. The larger the organization is, the more difficult it is to control and audit what is going on, and to correct mistakes in a timely manner. The reason that large companies remain successful is that they have economies of scale, financial strength, and often brand names. It would be wonderful to combine these advantages with the efficiencies of small companies where the decision-making is less cumbersome and there is less wasted time and effort.

Therefore, I advocate running large companies like several small ones. Some corporations adopted this view and consciously organize themselves that way by creating small business units (SBUs).

It takes great organizational skill to find the right balance between SBUs and their independence, and taking advantage of corporate size and might.

In order to understand how this balance can best be achieved, I will cover subjects such as span of control, organizing by function, corporate staff role versus line organizations, and auditing functions.

A business must have a purpose, a philosophy, and a reason for being in that business. The organization must reflect that philosophy. The organization should be able to change with changing times and conditions, but that should be an evolution (unless drastic measures need to be taken) because every reorganization effort brings some chaos and rivalries, no matter how well it is communicated. Large corporations have a habit of reorganizing too often, sometimes in order to satisfy the egos or aspirations of some of its executives. That is not what I mean by re-inventing yourself as a corporation.

I will describe various forms of organizations, and point out why some are better than others. Every form of organization is a compromise because different functions have different and sometimes contradictory optimum modes of operation. For instance, sales would like to sell everything possible, while accounting would only like to see sales that generate large profit margins. In any organization, there must be checks and balances, and in this case, someone must review all sales to ensure that they meet corporate objectives. That "audit" function cannot reside in the sales department.

How should a function be organized within the corporation?

There are some general principles that I will describe here.

Corporations should always be organized in a market-oriented fashion. Therefore, marketing consideration should come first and marketing should call the shots. No matter that the company has the best engineering talents coming up with new products, or that manufacturing just bought some expensive machinery that makes it the lowest cost producer, if the market changed and does not want your products you will not be successful. Therefore, decision-making regarding products and allocation of resources (i.e., the business plan) should reside in the marketing department. The product line managers in marketing should have wide-ranging powers to steer the organization in the right direction and to use the resources from other disciplines to accomplish their goals. Of course, there will be rivalry for resources among product line managers, but that should be resolved within the marketing department or by the CEO.

An example of mistakes made when a company is not market driven was a semiconductor company that wanted to market watches made from its own semiconductors. When it allowed key semiconductor personnel to plan, market, and sell the watches, it failed. The reason is that the people in the semiconductor company did not understand the watch market.

They should have set up a separate division for the watches, or at least a separate marketing and sales organization staffed by people skilled and experienced in the watch market. Watches are sold in different channels than semiconductors are; they require different people in sales and service, the business has different ratios of salespersons to dollar volume, etc.

Another example is that of a company manufacturing artificial turf. The company decided that it wanted to go into the business of installing the artificial turf because it could increase market share that way. It thought that the fact that its installation business competes with some of its customers was not going to hurt it. When the company staffed up for this diversification, it used some of its own people in key positions. It was unsuccessful in establishing and growing a service organization because its key people did not understand the market. The company would have been better off establishing an independent profit center and treating it like a start-up company being financed by the parent, and then hiring people who know about the turf installation market.

Being market driven does not mean that all functions should be organized along markets. Other functions should be organized along centers of excellence because if the talent pool is concentrated in one location, they can learn from each other, interface within their disciplines, be up-to-date with the latest techniques, and provide back-up talent when needed.

Engineering ideally should be organized by the type of engineering that is required. This allows for the best use of engineering talents. Engineering should be organized into specific engineering disciplines such as software, hardware, styling, etc. Naturally, some projects require a team from different disciplines to form a task force, but the home base for each engineer should be along his/her engineering functions.

On the other hand, manufacturing should be organized along manufacturing processes for economies of scale and efficiency. A corporation manufacturing various electronic products in various locations should not have PC board manufacturing in every one of those locations because the investment in capital and the economies of scale dictate a centralized manufacturing plant to produce all of those boards in one location (provided it is cheaper than outsourcing them).

Sales should be organized along regional territories to be close to the customers because that provides the best coverage for the market. Salespeople must be in close physical contact with their customers and that is best accomplished by regional orientation.

In a large global corporation with several divisions, every division will want its own "empire" of manufacturing, engineering, sales, service, etc. Even within each division with several product lines, each product line manager will want his/her own team of specialists in these disciplines to give him/her quick access and first priority. These desires are the exact opposite of specialization into "centers of excellence" within the corporation, which is what I advocate.

Even the location of each of these centers of excellence could be different for the corporation. Sales should be in locations to best serve the customers, while engineering should be where the skills required are available, and manufacturing should be located where labor for the type of manufacturing required is available at the lowest cost. Marketing and product management could be co-located with either sales or engineering offices. In today's world with easy information transfer and low shipping costs, it should be possible to organize these functions with a minimum of compromises into the best locations for each function.

Having "centers of excellence" concentrated in one location and under one manager inevitably leads to the need for a parallel matrix organization. That is what makes product line management such an important function. Every corporation, especially large ones, will have many different products. These can be grouped into product lines and managed by product line managers reporting to marketing. Product line managers are needed to give sufficient focus to a line of products. Product line managers will operate in a matrix organization mode, where they "subcontract" for the services of the various functions, located in their "center of excellence."

Where does this principle of "centers of excellence" fit into the small business unit (SBU) concept?

An SBU is organized to serve a certain market and thus it is a marketing-oriented organization. An SBU can have several product line managers if the line is broad enough. Each SBU has its only profit and loss center and its mini-CEO responsible for that business. That mini-CEO will be responsible for the business plan of the SBU and, as such, for the organization of the functions within the SBU. It is possible to have an SBU without any manufacturing as that could be subcontracted out either within or outside the corporation. An SBU is useful in order to divide large corporations into manageable profit centers, where responsibilities are clearly defined along with the authority to run a profit and loss center.

The SBU concept is helpful in keeping the span of control for corporate CEOs under seven direct reports (not including staff). One CEO can handle five or six SBUs.

Any large corporation will require a corporate staff. While essential to control strategic functions in the business, this corporate staff does not have to make a profit and has a tendency to grow and assume responsibilities that are best left to line organizations in the divisions. It is important to define what corporate staff should and should not be concerned with, in order to keep that staff "lean and mean."

Corporate staff should be an extension of the corporation's CEO. It should be designing and planning overall strategy and auditing the line organizations to ensure that the business plan is met.

Since they dictate and audit policy, corporate staff people must understand the business better than anyone else in the corporation. They must be the best-in-class people, rather than people shuffled aside by a reorganization.

Many corporate staff organizations employ people who fail elsewhere or are a parking place for retiring executives. That is not good policy and does not fit the idea that they are an extension of the CEO.

In some corporations, corporate staff concerns itself with trying to impose commonality within the corporation of small stuff, like printed forms or other details best left to the divisions. There could be some justification for using corporate purchasing power to reduce costs of materials, or some commonality of personnel policies across the corporations, but care must be taken to limit these corporate functions as much as possible because local people are better suited to deal with varying conditions and can be much more flexible and more effective than some corporate staffer far removed. In addition, corporate staff should be lean and therefore should not have time to deal with small stuff. Corporate staff should not spend time on "commonality" issues, unless the payoff is large. Remember, corporate staff do not have to live with the consequences of their actions or rules, whereas line people do.

The principle that authority should be given with responsibility applies here as well. The role of corporate staff should be made clear by a charter outlined by the corporate CEO. The role of corporate staff should be to review and audit the implementation of the various business plans, not to act as a "consultant" to the line organizations. It is up to the CEO to make sure that corporate staff does not keep growing like a tumor.

For example, should the corporate staff care that the signs on the buildings are uniform? Should they care whether the forklifts are all purchased from the same manufacturer? Should they care whether personnel policies and wages are the same across different states or countries? Should they compare cost of incoming inspection across different divisions? Should they care whether different divisions use different material control systems? If the answer to questions like these is yes, then you should re-examine their charter.

If the corporate staff cares only about the fact that profit and loss centers are making the profits according to the approved business plan and that they comply with the long-range strategic plan of the corporation, then they have the right focus.

Following are typical corporate staff functions:

- Help in formulating the business plans for the corporation.
- Audit adherence to the plans.
- Financial controls and audits.
- Allocation of resources between divisions.
- Human resources common benefit packages (but not individual divisional policies).
- Public relations, but again restricted only to common themes among all divisions.
- Purchasing contracts, but only large generic ones. Many times, this is more trouble than it is worth.
- Arbitration of transfer costs between divisions. A lot of wasted effort occurs in large corporations by arguing over transfer costs, when one division buys from another or there are joint projects or joint marketing budgets. When two divisions have conflicts, there should be a corporate arbitrator resolving these issues. Corporate staff is in the best neutral position to solve this. They may not like it because it is not a popular task, but they are best suited for this task.

7
Manufacturing Strategies

The reason for a manufacturing strategy is to look at manufacturing alternatives, and optimize the return on investment from them. Everybody wants to be the lowest cost, highest quality manufacturer using the latest technologies, but that is not always possible given budgetary constraints. In today's global economy, cheap transportation costs, fast moving technology, and just-in-time deliveries, it is mandatory to re-examine the manufacturing strategy each year and include it in the business plan.

Here are some of the considerations relating to manufacturing strategies.

OUTSOURCING AND MAKE OR BUY STRATEGIES

The most important manufacturing strategy decision facing management is whether to manufacture products or parts of the product "offshore." The term offshore in this context means to go to cheaper labor markets away from the home base, not to go across the sea. For instance, going to Mexico in this context is also called offshore. This term even applies to Asian companies in China or Korea, when they can go "offshore" to cheaper labor markets like Vietnam.

After deciding what to produce at home or close to home, the next decision is whether to make it or buy it. Manufacturing managers may want to make as many of their parts as possible or even advocate vertical integration because it makes their factory expand and their role more important. In the best interest of the company, these decisions must be made based on economic consideration. When make or buy decisions turn out to be in favor of buying, there will be plant managers who will not like that decision. The finance department often is the best one to analyze the problem and provide the right numbers, and recommend the action.

In order to make an intelligent decision whether to make or buy a product or part, only the variable overhead should be used for the comparison.

Some overhead costs are variable, some are fixed, and some are only fixed for a certain amount of time. (See Chapter 17 on how to calculate variable overhead.) The financial analyst is in the best neutral position to make the determination of what the variable overhead is that needs to be applied in each case.

Numbers are not the only consideration for make or buy decisions.

It is sometimes difficult to crunch numbers and come up with the optimum answers because there must be some assumptions made regarding vendor capabilities, pricing, and future trends. In offshore decisions and make or buy decisions, sometimes there are strategic considerations, like proprietary products, that override the numbers. These are:

- Considerations regarding vertical integration and the degree of automation
- The amount of capital available and the return on investment worthwhile
- As a general principle, it should be noted about outsourcing that a company should keep its core competencies in house

Once the decision is made on what the factory should produce, plans can be made on how to manufacture those products. Listed next are some of those strategic considerations:

- Large corporations that have several manufacturing locations should consider "focused factories" that make only a few products or specialize in a certain type of technology. Some of these considerations have been covered in the previous chapter where I talked about centers of excellence. For example, a large corporation manufacturing electronic products in several locations may not want to manufacture circuit boards in any of their assembly plants or system houses. They can either outsource all circuit boards or make them in a focused factory and ship them to their assembly plants.
- In some cases, factories do not have to be co-located with functions like sales or marketing, but in other situations, it may be desirable to co-locate some functions like engineering and manufacturing.

The strategic considerations on locations are how much interface is needed or desired between the disciplines.
- The manufacturing strategy should address whether the location of the factory or factories is still competitive. The location should have competitive labor available, including skill levels required. In case of a new factory being considered, a task force should evaluate alternative locations, tax considerations, energy costs, and a long list of requirements specific to the products being built. It can be useful to go through this exercise on a smaller scale to evaluate the justification for staying in place with an existing factory.
- The manufacturing strategy should also address whether the factory equipment and systems are modern enough to be competitive. This exercise should lead into having the right input into the business plan and the budget. Sometimes, in order to stay competitive, large capital expenditures are required, which may not be available.

CAPITAL EXPENDITURES

In our rapid technological change environment, it is important to "think big" and consider robotics, mechanization, and computerized systems that require large capital outlays and in some cases transform the operation. These strategic manufacturing ideas must be formulated to become part of the discussion preceding the business plan each year. Don't just give lip service or use traditional thinking when formulating this strategy. Robotics (or mechanization) requires a new approach to old manufacturing concepts that may not be apparent in the daily routine of cost reduction or continuous improvement.

The capital budget for manufacturing must compete for resources within the business plan. It is up to the manufacturing management to make the case for justifying large capital expenditures in order to modernize the factory. Only variable overhead should be used to justify these—or any other—capital expenditures. A financial analyst should determine what variable overhead applies in each case.

Even when capital equipment is justified and the pay-off is acceptable, there are limits to what the company can afford to spend each year, and priorities must be established to determine which proposal gets the green light in any given year. These priorities can be based on return on investment, but strategic consideration may also play a part in the decision.

MANUFACTURING CONSTRAINTS

In any growing manufacturing company, there are constraints to growth. Sometimes manufacturing has plenty of capacity and the constraint to growth is sales. Other times sales gets ahead and manufacturing is the constraint. We will deal here with constraint in manufacturing. When dealing with manufacturing constraints, you must find the weakest link or the bottleneck and fix it. After fixing one constraint, the next one should become obvious and you have to keep doing that until there are no longer constraints. Fixing bottlenecks is not new, but often the manufacturing manager should find them before they occur in order to anticipate a surge in demand. Capacity planning should be the way to do that.

The capacity of a specialized very costly machine could be the bottleneck in a factory. There are several solutions to be looked at before deciding to buy another costly machine, which the company may not be able to afford. The obvious solution is second shift, and even third shift, working on weekends, etc. but there are other not so obvious possibilities. After exhausting 24/7 possibilities, manufacturing engineering ought to study the operation on the machine that is creating the bottleneck and consider the following:

- Add a second operator to feed or to unload (even if it seems inefficient).
- Examine the downtime and how it can be eliminated. Quicker changing of dies, quicker feeding or downloading, and examination any time the machine is not operating at full speed to see how that can be mitigated.
- Examine if machine operators cause downtime by taking breaks (utility person should be assigned to step in), or downtime between changing shifts, etc.

If all of the above are in place and the schedule is met, examine what the capacity is after all steps are taken. If capacity is still close to 100%, then you should have a plan in place in case demand is increased.

However, all that may not be enough. If you have a press that is continuously working 24/7 or close to it, what happens if that press breaks down? How long does it take to have it operational again?

Are spare parts easily available, and should you be stocking them? Some large presses require several months to refurbish and a new one may take 6 months to order and set up. What is your plan if that happens? If that is the case and subcontracting the operation is not a good option, maybe it is better to buy a second machine, even if both machines do not operate full time. As you can see, breaking bottlenecks is not always as easy as it first looks.

If the cost of another machine is prohibitive, redesigning the product to avoid this bottleneck or subcontracting some of the volume could be the solution. **Never be constrained by manufacturing capacity. Always plan ahead to break the next bottleneck that will occur when production needs to be increased.**

There can be constraints other than sales or manufacturing. Time to market being too slow can be a constraint for a technology company. In that case, the corrective action is to review the process of new product design and new product introduction and speed up the process.

An interesting example of manufacturing constraint occurred when I was hired to take over management of a Nortel factory in Silicon Valley. Canadian managers ran the factory until I was hired. They hired me because they wanted a fresh perspective because performance of the factory was substandard and I was a product of Silicon Valley. Not only was the factory losing money, but also it could not deliver a surge of orders. When I came, I discovered that the bottleneck was testing our circuit boards. We had modern equipment, but it had to be programmed and we did not have enough test engineers to program it. Upon further examination, not only could we not hire enough test engineers, but also we had been losing the ones we had. We needed 47 test engineers and had a dwindling number of 25. I discovered that we did not pay them enough to keep them, let alone hire new ones. The reason was that our pay grades were established by our Canadian parent company and Test Engineer was a pay grade 7, which was not enough to compete in the market in Silicon Valley. The Human Resource Department would not allow exceptions and the Canadian General Managers before me would not pick a fight with the parent company. One of my first acts as General Manager was to reset the pay grades for Test Engineers and Test Technicians to 75% of the average Silicon Valley wage rate. One of our competitors was Rolm, which had swimming pools and exercise rooms for employees, free coffee and donuts, and other amenities. I decided to shun all this and competed on wages alone, and selectively set a higher percentile for those skills than our competitors did. Within 6 months, we had the required 47 test engineers in place and never had a bottleneck again in that area. The factory delivered on its promises and became profitable again.

The lesson learned here was that wages are more powerful incentives than benefits and if you desperately need a special skill, it is best to pay above the going rate. In Chapter 11, I talk about how to set standards of pay.

COST REDUCTION

Most of the money to be saved in manufacturing companies does not come from hourly employees working faster. It comes from reviewing specifications, redesigning parts, changing vendors, or changing manufacturing strategies. There will be strong objections in doing any of these, from quality, from engineering, and from materials. That is why the general manager should include in each department head's MBO (management by objectives) the reduction of manufacturing costs by a certain percentage. As always, the amount of cost reductions should be an attainable goal, tasked to be done by a cooperative effort, where an interdisciplinary team should be jointly responsible to meet those goals. Otherwise, why should engineering work on redesign, when designing new products is so much more fun? Why should quality bother to take a chance in relaxing some specifications and take time having to run tests? Why should materials look for other vendors or outsourcing, when they are so busy with current problems?

The answer is because there is a lot of money that can be saved for the company in cost reduction.

Therefore, I believe that a permanent task force is needed if the company is serious about reducing costs. The reason for such a task force is to prioritize where to invest the time and money to achieve the most cost reductions. This prioritization is necessary because there are many different ways and ideas to reduce costs but they cannot all be done due to the efforts required by various departments to achieve it. That is why prioritization by a task force is important. I believe in an interdisciplinary task force because the largest cost reductions come from ideas that cannot be implemented without consensus by several departments. Manufacturing may want to ease up on some specifications, but that requires engineering and quality approval. There may be a proposal to make engineering changes to facilitate manufacturing and that requires engineering priorities. Purchasing may need to evaluate different vendors or use alternative components that could save a lot of money, but they need the approval from engineering, quality, and manufacturing. All of these approvals may take some time to investigate and longer to implement. That is why the different disciplines must be represented in the task force. The task force should be chaired by the manufacturing engineering manager and should include the department heads of materials or purchasing, engineering, quality, and finance. Unless the department heads are involved and attend these meetings, there will not be enough commitment on the part of these functions to make the cost reduction happen and to make the necessary decisions. Is there enough money involved in cost reductions to take up the time of these department heads? You bet there is!

After deciding what to work on, the chair of the task force should assign the work to be done with target completion dates and run the cost reduction meeting like any other project meeting with actions assigned and progress against them being reported at every meeting.

Like any worthwhile program, the cost reduction task force should have the backing and support of the CEO, it should have annual targets, and the finance department should report actual savings versus the target. In this case, savings should be calculated using only variable overhead because when you are cutting costs, the fixed overhead remains the same.

Value analysis is required to reduce costs of an existing design. While the task force is comprised of department heads that do not have the time to do that, they should delegate this to those most qualified to come up with this analysis and present it to the task force. As a side note, a typical example of value engineering is that of a tie clip costing $5. The same result can be accomplished with a large paper clip costing a fraction of pennies. Of course, this is an exaggeration and only works if the tie clip is not used for decorative purposes, but it illustrates how far value engineering can reduce the cost of a component.

Having a cost reduction task force does not mean that this is the only way to reduce costs. It is the task of manufacturing engineers to make sure that labor, materials, tooling, and systems are optimized. This optimization should lead to reduced costs of existing practices without large-scale changes necessary. A good rule of thumb is to have a manufacturing engineer for every 100 workers in a factory.

MATERIALS SYSTEMS AND SUPPLY CHAIN MANAGEMENT

The materials function is responsible for purchasing materials, scheduling the factory, storing and delivering materials to the factory, and shipping the product to the customers. Material systems should be integrated so that one transaction triggers as many data transfers and transactions as possible automatically. For example, an integrated system will enter an order to the system and then automatically generate all work orders and schedules necessary to manufacture the product and all paperwork necessary to ship the product. One good way to reduce transaction costs for items that do not cost much and have no value when stolen is to eliminate transactions by making them free stock on the production floor.

The materials department controls inventory in raw materials, work-in-process on the shop floor, and finished goods. They rely on the forecast to generate requirements for the inventory. There is constant striving to minimize inventory because it consumes capital. The less inventory the better from an asset management or any other point of view. On the other hand, sales can be lost if the company cannot meet the demand, whether or not it was forecasted. Modern systems rely on the supplier chain and just-in-time systems (both for vendors and in-house Kanban-type systems) to meet these seemingly conflicting goals. Many factors enter into the decision on how far to carry these concepts. There are obvious risks attached to just-in-time vendor systems because you are relying on someone else for quick reaction times and ensuring that the capacity exists to fulfill your orders just-in-time. There are distance problems with global manufacturing where the lowest costs are from distant shores.

In any manufacturing company, it is necessary to have a material control system to ensure that the right material is delivered to the right place at the right time. The system to enable these objectives ranges from the old-fashioned MRP system, to the ideal just-in-time system. As we discussed previously, just-in-time is not always possible, but it is a goal to strive for. What is often forgotten is that not everything is covered by these systems. Supply items or spare parts for machines can also cause delays or stoppages in production, especially in countries with little infrastructure.

Some material organizations in which I was involved worked on the "SWOT" system for these supplies or spare parts. SWOT stands for "Shit, We Are Out Again." The cure for this sickness is to institute an informal min-max system. A min-max system establishes a minimum level and an order quantity for every "free" item. There are many ways to trigger the order when the minimum level is reached. One of the simplest ways is to package the minimum quantity so that when that package is opened, an order is triggered.

The problem with min-max systems is that as time goes on, the quantities can change or even become obsolete. Therefore, with any min-max system it is necessary to have a list of what is included and review and revise that list every year. It is also important to add new items to the list and to the stock as new products are released.

Materials management has to take some calculated risks and must be smart in supplier chain and vendor relationships. They must work very closely with sales to ensure that forecasts are updated often and with the right input. They must create reports to keep on top of balances in the system. It requires constant vigilance to ensure that products are delivered on time and that inventories are kept low. Most importantly, they must develop vendor partnerships. The old-fashioned adversarial relationship with vendors has no place in today's modern manufacturing management.

To give you some perspective, the old time management approach for reducing costs and reducing inventories was as follows:

- Beat down the price to the lowest possible priced vendor.
- Demand that the vendor carry your inventory and deliver it just-in-time.
- When your forecast does not meet expectations, cancel orders, or reschedule the vendor.
- Develop two sources for every major part or component.
- Do not share your schedule or forecasts with your vendors.

All of the above tactics were tried and failed to produce acceptable results. It is easy to see why.

Vendors who operate under the aforementioned conditions cannot produce efficiently, cannot invest in reducing costs, and eventually fail to deliver quality products on time. When you second source, you create more vendors instead of less, and you may end up with two substandard suppliers instead of one good one. Your relationship costs with second source vendors are double than they would be with single source ones, and these relationship costs are substantial if you manage your vendor relationship correctly.

When you destroy most of your vendors and only one or two survive, they have you over the barrel, rather than the other way around.

In the modern world of supplier management, it is important to consider your larger vendors as an extension of your factory. This strategy of close cooperation can result in just-in-time deliveries, sharp reduction of incoming inspection, and continuous improvement in the cost of products from your vendors, as they can rely on you for continuity. It also fosters vendors who are fiscally stable and will stretch their resources and brainpower to help you in your business.

The tactics used are called supplier chain management and they range from having buyers at supplier plants, to sharing cost savings and long-term contracts.

Here are a few principles of supplier chain management:

- Most important is the philosophy that your vendors are an extension of your factory and not "the enemy."
- Choosing single vendors that have the capacity and willingness to partner with you for large orders or key components.
- Sharing information with vendors regarding your business and demanding that they also cooperate with your people by sharing how their costs are developed.
- Working with your vendors' staff at all levels to understand each other's concerns and problems. Working at all levels means not only purchasing, but also scheduling, quality, manufacturing, and engineering, even cost accounting.

Don't think that these tactics mean that your purchasing people should not look for lowest cost and coddle your vendors. If you understand more where the shared costs are, you will arrive at the best compromise.

Supplier chain management works best when your business is a large percentage of your supplier's business. It is naïve to think that if your purchases from a vendor consist of less than 1% of that vendor's business, you can dictate terms to the vendor. Using your common sense, you should realize that your vendor to whom you are unimportant would not change the way it does business. Does this mean that you are powerless to control quality or price? Not at all! When your vendor is much larger than you are, instead of dictating procedures to the vendor, you must make it easier for the vendor to do business with you. You have to work at lower levels of your vendor's organization to get cooperation. You must become proactive in understanding how your vendor manufactures the part that you are buying, and how you can influence the outcome. If the part that you are buying is a substantial amount of the cost of your product, you must understand how it is manufactured at your vendor and what you can do to help the vendor reduce that cost and ensure a quality level that eliminates your incoming inspection. Sometimes that is not possible, but it is worthwhile to try. Sometimes you must change your designs or specification to buy the vendor's standard part and accept the vendor's procedures for deliveries. Your manufacturing and quality engineers should visit the vendor working with their counterparts to come to an understanding on how to achieve your goals. In some cases, your financial people can get involved also by extending longer-term contracts to get a lower price.

It is not always the best strategy to give all your problems to your vendors or to have just-in-time deliveries. For example, suppose there is a factory in a rural area with cheap land and ample warehouse space left over from discontinued products. Suppose that factory uses many cardboard boxes that take up a lot of space and the boxes must be pre-printed, which takes expensive set-up time. Therefore, just-in-time is not a good option for the boxes. The box vendor has a just-in-time program that works well for most vendors who do not have much warehouse space, but it charges 12% per year for just-in-time deliveries because he must store the pre-printed boxes. The rural manufacturing company would be better off if it would take larger box deliveries based on economic order quantities (may I be forgiven for using such an old-fashioned term), and store the boxes in its warehouse, rather than force the box manufacturer to have frequent set-ups or store some of the material for later just-in-time delivery to you.

OFFICE LAYOUTS

The latest trend in office layout is to have open bullpens, with several people sitting around the same area interfacing with each other. The only offices in this scheme are small conference rooms for occasional common use. This may be ideal for high technology companies where a lot of creativity is required, but it should not be adapted for low technology manufacturing companies just because it is the latest trend. Many clerical and staff functions do not require frequent interfacing with others, and people are more productive when not disturbed or distracted. They need cubicles or partitions. When laying out offices, the type of work done in them should be considered, and common sense applied as to what leads to more productivity and less errors, rather than following the latest trends.

While I believe that open and shared office layouts are not for everyone, I also believe that all production support personnel, including quality engineers, manufacturing engineers, production control materials personnel, and production supervision should be on the production floor as close to the action as possible.

8

Manufacturing

Production should be organized into work centers based on processes or products. There are two schools of thoughts about how a factory should be laid out: whether to lay it out by process centers or by product flow, which integrates different processes and machines into a cluster.

The definition of a cluster—sometimes called work cell—is that regardless of the process or machinery used, the layout allows the product to be manufactured in a continuous flow through the cluster. For instance, a punch press used in that flow will be located in the cluster, rather than in the press shop.

The definition of process-oriented layout is that similar processes are co-located in one center, regardless of the product flow. For instance, all machining is done in the machine shop, all plating in the plate shop, etc.

Often a mixture of clusters and process-oriented work centers is used. For example, if there are many products or subassemblies that have the same operations, it is wise to create a cluster of machines and tools that allows these products or subassemblies to be done in that cluster, while the rest of the factory is using process-oriented work centers. The cluster may have machines or processes that normally would be placed into a process-oriented factory. **The cluster reduces work-in-process inventory, material handling, and the number of transactions, and is a very efficient way to manufacture. Therefore, it is the preferred way to lay out a factory.** However, in many situations especially where the movement of material between operations is not uniform, it is not possible to use clusters. In that case, the factory should be laid out in process-oriented work centers. The advantage of the process-oriented work center is that operators can be cross-trained and back-up machines are available when needed.

The disadvantage of process-oriented work centers is that material is moved from one work center to another creating larger than necessary work-in-process inventories and slowing down lead times. Of course, despite all these disadvantages, all plating will probably be best done in the plate shop, rather than scattered about the production floor. The ideal layout of a factory uses clusters with product-oriented flows, and has little or no work-in-process. The work flows smoothly like on an assembly line, regardless of the process required. Everything is simplified, transaction costs are minimized, and production is most efficient. Problems are caught earlier, there is less reject and waste, it is the ideal workflow. The ideal factory also is flexible enough to produce parts and assemblies custom built, or built in short runs, rather than making the same products in long runs. The reason is the same as previously stated: reduced inventories, less rejects piling up when things go wrong, and faster through-put in the factory. The layout of the factory should reflect these strategies, allowing little room for waste and shortening distances between operations.

Modern factories strive to come close to these ideals. The problems that sometimes cannot be overcome are the cost of duplicating equipment, short runs versus long runs that have different assembly sequences, and specialized equipment or test stations that cannot be put in-line with the flow of materials.

Sometimes it is best to have a separate "job shop" environment for short runs in order not to clutter up the factory with problems of expediting short orders that need special skills.

In this job shop, skilled operators build products from engineering drawings and custom build or modify special short orders. There is no need for manufacturing instructions or visual aids. This work center operates on a job shop accounting system. By making it a separate work center, the rest of the factory can operate on a continuous flow production and accounting system.

Production workers must be given tasks suitable to their skill level. It is ideal to break tasks down to small elements that are easy to learn, but that is not always possible. Some of the latest trends form work groups and self-supervised teams, but these experiments did not get any traction for very good reasons; namely, lack of supervision and too many complicated tasks. Don't fall for the latest fads coming out of high-technology companies that have unusual production situations or are making a lot of money due to their marketing position and can afford social experiments.

In 1987, Volvo embarked on a social experiment to build cars in a revolutionary way. They built a new factory (Karman) but instead of an assembly line, they adopted a team assembly system, where the team had to have enough skills to perform multiple operations and complete assemblies on a stationary car. The idea was to relieve workers of highly repetitive tasks and the drudgery of a moving assembly line. Workers were happy until Volvo had to close the Karman plant in 1994 because it could not compete with others that used assembly lines. After that, those happy workers were laid off. Volvo was later sold to Geely Holdings, a Chinese company that is now operating this Swedish firm. This sad story speaks for itself. There is no room in manufacturing for social experiments.

Most high-volume production uses assembly lines or tasks that require minimum skills. It is better to provide a lot of support to production in the form of manufacturing engineering and materials, rather than task production employees with diverse operations. The simpler and more repetitive the task, the easier it is to measure the output and to train the workers.

The best way to ensure high productivity is the assembly line. Assembly lines were abused in the 1980s especially by the U.S. automobile industry by not allowing workers to stop or slow down the line when problems arose. This has been corrected and is now standard practice (if it is not in your factory, it should be) that anyone working on the line can stop it when something goes wrong, a part is missing, or it is going too fast to keep up with the work. An assembly line does not have to be moving; some manufacturing operations adopt a push system where each operator passes or pushes the assembly to the next one. The important thing to keep in mind when setting up an assembly operation is to make sure there is a continuous flow, and break down the job so that each operator can perform his/her task, given simple instructions or visual aids. The greatest drawback to assembly lines is that if an operator is absent, there must be a substitute to take his/her place. If that substitute does not know the operation, it will slow down the whole line. Therefore, it is wise to train a utility operator who knows all the operations and can substitute for an absent operator.

During Detroit's problems with quality, it was well known by insiders not to buy a car that was assembled on a Monday. The reason was that there was more absenteeism on Mondays than any other day, and the hard-headed "tough" management did not want to lose any production because of that. Since assembly lines were very long, there were not enough people to operate the lines, but management did not allow any slowdown because that would mean getting less cars out at the end of the day. Cars ended up being assembled without some parts and with some untrained people who could not keep up with the assembly speed. This resulted in bad quality that could not be remedied.

The manufacturing organization must deliver quality products on time. This requires discipline and discipline requires close supervision. To effectively supervise production employees, the supervisor needs the tools for measurement and feedback. That is the secret for productivity on the shop floor, and in order to give the supervisors the necessary tools they need training and they need standards of performance and measurements for every task, or as many of the tasks as possible. The best way to evaluate a production operation is to look at the number of operations that are measured and reported against. The amount of work supervisors do should be kept to a minimum or, in most cases, zero because the supervisor should use his/her leverage to make sure that the employees are productive. If the supervisor performs operations, that is a sign of poor organization where there are too many supervisors for too few people, or where the supervisor is poorly trained.

The number of employees a supervisor can oversee depends on the skill level required for the operations. If the supervisor is smart and uses leverage, he/she can appoint working leads and thus increase his/her span of control. A bad recipe for floor supervision is to have a supervisor for a small number of workers and then have him/her do some work because he/she will have time left over from supervision. For instance, if you have a production line with eight people, you don't want to have a supervisor for them, but rather a working lead. A production supervisor should have 30 to 50 direct employees with working leads as required.

When I worked in Mexico and set up production, we had to promote some operators to become supervisors. Do you promote your best worker to be a supervisor? It turned out that when we have done that it was usually a mistake. When we observed a group of people at lunchtime or at group meetings to see who was their leader, or who was the most outspoken, and promoted him/her to supervise the group, we had much better luck with our selection. This may be a good lesson on promotions. It is not always the best worker or the best salesperson who makes the best supervisor, but rather the one with leadership skills.

In a manufacturing company with steady product lines, the manufacturing costs should be going down every year as they get more efficient, as productivity goes up, and as they find smarter ways to manufacture the product and buy parts. It should be recognized that people do not cut costs willingly and it requires effort to reduce costs.

Part of that cost reduction comes from involving workers on the factory floor in helping to make production more efficient and improving the quality and yield. Many programs address this issue and, in my opinion, quality circles is one of the best.

Following are some of the most popular manufacturing programs.

Six Sigma was invented by Motorola with the goal of guaranteeing 99.99% perfection, hoping to get as close as possible to that lofty goal. It is not a bad program because it raises awareness of quality and therefore it is useful. Certainly, Six Sigma has its uses; for instance, if you measure process variations on a lathe where parts are turned out with tight tolerance and you want to know when to change tools, you would use Six Sigma methods. The Six Sigma method is basically a statistical control tool to ensure not only that quality is good but also that is remains good. Your manufacturing engineers should be aware of this discipline and use it where it applies.

Total Quality Management (TQM) and "Zero Quality Program" have been useful programs for factories that could not get their quality to measure up to customer needs. They have ideas that are useful and sometimes essential to produce a quality product. Here again a good quality control plan is needed in every manufacturing company and that can be accomplished by using these methods. I object to the term "Zero Quality Program" because it sets an impossible target and challenges people to come close to it. I believe the champion for TQM should be the quality control manager and he/she is the only one that needs training in its principles and then he/she needs to implement it in the company rather than everyone else being trained in a company-wide program.

Lean Manufacturing started with Toyota as a system of eliminating waste and has some very good principles incorporated under one umbrella name. If the company adopts this slogan for continuous improvement, I have no quarrel with that as long as it is understood that it is not a cure-all for manufacturing inefficiencies and it is a top-down system. Here again, the training should be done and restricted to manufacturing engineers, who are then responsible for running the program.

Quality circles started in Japan and while very popular in the 1980s, it had limited success in the United States. Unfortunately, it was misused as a management tool instead of a bottom-up involvement of production workers in problem solving.

Quality circles are not meant to be a top-down management program.

It is a good system for getting people on the factory floor involved and therefore it is a useful motivator. I personally like the idea. My preference is to call the program productivity circles, but only start it if you fully understand the implications.

It involves 8 to 10 people in a section on the factory floor, given training in problem solving. They meet during working hours to discuss problems of quality and production using the problem-solving techniques they learned. Meetings are held in an organized way every week or every month. The program is run by the group under a leader who either is elected from the group or is a lead or supervisor of the group. Participants are encouraged to bring quality or cost problems to the attention of the group. The group then prioritizes the problems and engages in problem solving, based on the training they received in that discipline. Some form of recognition by management or human resources is essential to reward good solutions that have been implemented. This program has many advantages. It enlists the workers on the factory floor to care about the quality and cost of what they are doing. It shows people that the company cares about what they are doing. (See the Hawthorne experiment about the effect of someone watching and caring.) It gets ideas from the people doing the job, which they know more about than anyone else. It is not a top-down program. An advantage is that it frees up the most valuable resources of the company to be engaged in higher-level activities, rather than troubleshooting and problem-solving minor issues, and it is a self-administered program.

All of the above require effort by management to implement it and to train people to do it. One system does not do it all. Excellence in manufacturing requires all of the above. This should not be scary; all these "buzzwords" consist of nothing else but good practices of managing a manufacturing company, packaged in ways emphasizing one aspect over others. Consulting companies sell their services to implement these systems (and many others, like inventory management, supplier chain management, etc.). I believe that in a manufacturing company the champion for efficiency and cost reduction should be manufacturing engineering. The champion for quality should be the quality control manager. The champion for training should be human resources. If all these managers are doing their jobs well, you don't need consultants. If they don't, consultants will only accomplish temporary results and things will revert back to substandard performance. I recommend that managers study the latest trends and go to seminars that teach them, but they should make up their own minds on what applies best to their operations.

Some mention must be made of manufacturing systems that are imposed upon the manufacturers by vendors wanting to ensure compliance to good practices. One of the worst is ISO 2000 and its derivatives, sometimes called ISO 9000. ISO stands for International Organization for Standardization. ISO 2000 is supposed to improve and certify the quality of the product by making sure that procedures cover every aspect of manufacturing and quality. That program was imposed upon manufacturers by large vendors, who were trying to ensure that the product they received was free of defects. Procedures will not do this. Compliance with the program has been taken over by consultants who write the procedures, certify them, and then leave. They usually guarantee that if hired the certification is assured. Companies that hire these consultants routinely ask them to leave their personnel alone, give them a small office to write their procedures, and after that bid them good bye and good riddance until next year when they have to re-certify. Chinese consulting companies have sprung up who understand the language of ISO 2000 and sell their services to Chinese manufacturers who know nothing about ISO 2000 and whose employees cannot read the procedures. There is no auditing, no follow up, and no guarantees that the product gets any better from ISO 2000.

Any buyer accompanied with a manufacturing engineer or a quality engineer with a checklist and a one-day visit to a factory can evaluate the ability of that facility to supply a quality product far better than 10 ISO 2000 certificates.

9
Quality Control

Quality problems are one of two top reasons companies fail (the other one is running out of money). It is well known by now that no company can survive for long—let alone grow—unless the quality of its products is as good or better than its competitors. **Slogans will not produce better quality. A quality plan will!**

In a manufacturing company, the quality function should not be reporting to production. That is like having monkeys in charge of peanuts.

Some companies appoint a quality director to report to the CEO or to their corporate staff. That is not good practice in my opinion. Quality control is a line function that works within operations. If the quality director reports to the CEO, it is usually in name only and in fact, he/she is a corporate staff person. To have that function in corporate staff certainly ensures its independence, but it leads to conflicts and tensions within operations. The quality director cannot be a staff person. He/she has several direct reports and has line responsibilities (like in-process and incoming inspection). He/she not only designs the quality plan but also implements it (thus, the line function).

It is best for the quality director to report to the VP of operations as long as the company and the VP of operations is indoctrinated in a culture that believes that quality cannot be compromised. I don't believe that a VP of operations should be looked upon as a production manager. If he/she is not indoctrinated in the importance of quality, then no type of reporting relationship will cure the problem of him/her not understanding that quality cannot be compromised.

In the final analysis, it is the responsibility of the VP of operations to ship a quality product and any quality problems are his/her responsibility to fix. Therefore, along with responsibility should reside authority.

The quality plan should be an auditing plan where quality should not be the policeman, but rather the auditor. I will repeat it because it is so important to implement this principle in every facet of the quality plan. **Quality control should audit, not inspect.** It is wasteful for someone to produce parts and have someone else inspect every operation. It is wasteful to have trained quality personnel do 100% inspection, which should be the job of a relatively low-skilled operator.

The same theory applies to incoming inspection. They should not be inspecting incoming parts with 100% inspection. The quality plan should be based on audits and sampling. If the sampling plan shows a problem with the batch, that batch should be sent back to the supplier. If there is no time to send it back, quality should borrow an operator from production and charge the sorting time back to the vendor. This way the trained staff of incoming inspectors is relieved of repetitive work.

Desired quality level must be determined according to the product manufactured. A certain percentage of failures can be acceptable to most manufacturers. This should be articulated so that the quality plan will strike the right balance between 100% inspection at several points and random inspection at each level, with only final testing or final product inspected or tested at 100%. Even then, it is not shameful to admit that some amount of defects will slip through because not every parameter can be inspected and people make mistakes. **Perfection is a pleasant illusion.**

The quality level of an airplane manufacturer must be fail-safe, but the requirement for a plumbing manufacturer does not require the same 100% quality.

The quality plan should be an annual plan just like the business plan. It should deal with all of the facets necessary to achieve the desired quality level for the products. It should contain a quality calendar for the frequency of auditing processes and vendors. Some audits are required daily, some monthly, and some only every quarter.

Quality is a function that needs constant education of the workforce. The workers come from different companies and different cultures, with different education levels and different ideas. How can they understand what quality level is required and how to achieve it in their new work environment? I know the adage "quality is free," but that is not true. It requires extra effort. Production managers must be trained to allow delays or stoppages when quality is not up to standards. New employees must be trained in what constitutes quality work.

Quality must be built in. It cannot be inspected into the product.

The quality educational program for the company should include training programs explaining why quality is important, examples of critical processes, and explanation of the quality plan.

There should be a quality reporting system to measure progress for key quality indicators.

Procedures are more important in the quality area than all others. Corrective action formats and feedback must be part of the system. The quality function must be able to overrule production by stopping production or shipments until corrective actions are taken.

There are many tools in a good quality program, such as batch controls, sampling plans, first article inspection, etc. These tools must be fashioned to the individual process and cannot be detailed in a general sense here, but I will describe some general principles of good practices in a quality plan.

It is most important to specify what constitutes good or acceptable quality at every step of the operation. If these quality parameters are too tight, it could end up costing a lot of extra effort and money. The quality parameters must be developed by the quality organization, often in cooperation with engineering. You should not look for perfection, but rather look for acceptable quality.

Cosmetic appearance defects are difficult to define. **The principle of largest acceptable tolerance should apply.** If you examine a razorblade's edge using over 50× magnification, it looks like a lot of hills and valleys. With visual inspection of scratches or other cosmetic defects, you should specify a distance from which the part or article should be examined and define acceptable versus reject as viewed from that distance.

For example, a cosmetic specification may be to view a plastic cover from a distance of 3 feet and specify the visible number and length of scratches that are acceptable.

Visual aids are often helpful to distinguish between good and bad quality. Beware of the ones that classify appearances as good, acceptable, or rejects. You don't want to make a distinction between good and acceptable quality. You just want people to know what part is a reject and what part is not a reject. If it is not a reject, it is good.

In the electronics industry, we used to have pictures of solder joints classified as very good, acceptable, and reject. What is the purpose of distinguishing between very good and acceptable, when both are going to pass inspection? It only confuses inspectors. There should be only pictures of unacceptable quality on the visual aid. Everything else is good.

On a production line where the product passes through several operators, the most desirable quality plan is to have each operator glance at the quality of the previous operation before tackling his/her task. Random sampling by rowing inspectors would complement such a scheme, rather than having an inspector 100% inspecting at the end of the line. The last operator should have an operation that takes less time and allows him/her to inspect the assembly. Naturally, there are some tests that are needed to be 100% tests (like leakage testing or electrical tests), but these should be done by a production operator. Quality should not do anything that requires 100% handling or inspecting. The quality personnel's job is auditing.

Incoming inspection is a quality function in order to ensure that every part entering the factory is of acceptable quality. Note again that the emphasis is on acceptable, rather than perfect. The best tool for an incoming inspector is to have a specification or drawing, outlining critical dimensions and critical criteria. Here, again, the maximum acceptable tolerance variations should be specified. If the company holds its vendors responsible for quality standards that are too tight, eventually the company will have to pay the extra price for that quality level.

The principle of sampling applies to incoming inspection even more so than inside the factory. There are tools to use like sampling plans that can guide this activity.

In the global economy, parts or subassemblies can come from many different vendors from different countries. The supply chain and many just-in-time schemes force close cooperation between you and your vendors. Ideally, all inspection to acceptable quality standards is done by the vendor, and instead of incoming inspection, you audit the vendors' processes. That is the ideal situation and you should strive to get there for each vendor with which you are dealing. You should also understand that this is not always possible and cannot be implemented for every vendor.

In a perfect world, there would be no incoming inspection and parts or subassemblies would be delivered directly to the production floor. You should strive to get as close as possible to this perfect world on as many parts as possible.

The quality plan should include feedback of the audits to both production people and vendors. The type of feedback cannot be specified here because each situation is different, but after feedback there should also be a form for corrective action if that is necessary, and no further manufacturing or receivables should be done until the corrective action is resolved.

Programs for manufacturing quality sold by consultant companies address the same issues that have been described previously. Some of the best-known names are Deming, Juran, Total Quality Management, Six Sigma, etc. They serve a useful purpose because without a formal quality program there will not be assurance that every aspect of quality is well taken care of. It is up to the quality control manager to decide whether there is a need for a consultant or whether he/she can implement a good system, based on the guidelines here. Often it would save a lot of money if the quality director would just read up on any of these programs, or attend a training program and then implement it as part of his/her quality plan. You don't need to involve many people in the company to do the job of the quality professionals that are on board.

The quality director needs the support of the CEO and the training programs by human resources to have successful implementation and maintenance of programs. Sometimes, the quality director can run the training program better than human resources can and if that is case, it is preferable. I also suggest that the quality director chair a quality council to meet every quarter on quality issues with other department heads to get feedback and to discuss what support he/she needs from them.

The following table is an example of a quality calendar's top page, which should be followed by a timetable issued by the quality control manager.

Quality Calendar

Activity	Frequency	Months
Quality council	Quarterly	Jan, Apr, July, Oct
Education program	Semi-annual	Feb, July
Audit store fifo	Semi-annual	Jan, June
Audit self-inspect	Semi-annual	Feb, July
Audit procedures	Semi-annual	Mar, August
Audit calibration	Semi-annual	Apr, Sep
Critical parts review	Semi-annual	May, Oct
Quality awareness	Semi-annual	June, Nov
Review in-process	Quarterly	Feb, May, Aug, Nov
Review sampling plan	Semi-annual	Jan, June
Source inspect vendors	Semi-annual	Feb, July
Inspect tooling	Quarterly	Mar, June, Sep, Dec
Issue quality report	Monthly	Jan–Dec

This listing of activities is by no means complete. It is only a sampling of general topics, and each company should have its own activities list.

Every one of these activities should have a procedure and checklist, have a person assigned to execute it, and report completion to the quality council.

In the quality calendar, a date will be assigned to each activity to formalize when in the month the audit or activity is scheduled to take place.

10

Controlling Overhead

A large portion of manufacturing costs are overhead costs. One of the key jobs of manufacturing management is to keep these costs as low as possible. The dumbest way to measure overhead costs is to benchmark them against competitors or companies in similar businesses. The second dumbest way to measure them is to show them as a ratio of direct versus indirect labor. Here are the reasons why these comparisons are meaningless.

Different companies consider overhead differently. In some companies, quality inspectors and lead-girls are considered overhead, whereas in others they are considered direct labor. Some companies outsource more than others do, some have more automation than others do, and thus their overhead ratios are different.

To carry the argument further, in a lights-out factory that is fully automated with one technician (overhead) overseeing the operation, factory overhead ratio of direct versus indirect is infinite; yet, with infinite overhead it is the most efficient factory that could be desired.

How then should we measure the effectiveness of overhead and determine how much is too much? The answer is zero-based budgeting.

You should identify and then justify each overhead element and cost. To oversimplify, **if you can't put it in a box and ship it, don't do it.** Naturally, oversimplification does not work in real life, but the principle should be kept in mind.

Examine the job of each person considered overhead, and justify his/her existence. It is useful to classify what jobs are considered overhead. The definitions of direct labor and indirect labor are a good start, with indirect labor being overhead. A working supervisor, for instance, should be classified as direct labor if his/her job includes moving material because he/she works on the product. A technician repairing production machinery should be considered indirect labor. A supervisor whose job does not include working on the product, but is mostly managing others, is indirect. A maintenance person who maintains the facilities should be considered indirect.

Before tackling specific areas for the reduction of overhead, the company strategy should be to simplify manufacturing so that overhead should not be needed. This is done by simplification and elimination of bureaucracy. This includes fewer suppliers, reduced parts count, focused smaller factories, smaller plants, shorter distances, fewer reports, more frequent deliveries, less warehousing, fewer job classifications, etc. Reduce work-in-process (WIP) inventories. Counting, storing, and expediting are all overhead that costs money without contributing to the product. After you have done all that but not before, can you start your zero-based budgeting of your overhead. The reason is that if your overhead strategy of the above items is wrong, you are working on the small stuff instead of the bigger impact items.

Do not build rooms within rooms. When something goes missing and inventories do not match book value, accounting usually forces manufacturing to exert more control. They respond by building a fenced area with transactions in and out being counted and reported. They build a room within a room and now they have control. True, but control comes at a cost. Sometimes this is unavoidable, but often there are other ways to make sure that your inventory data are correct. If your WIP is small, if your production process is continuous and cycle times are fast, then there is no inventory to worry about. In that case, you should use the "four wall inventory system." Enter all the materials from the Bill of Materials for each assembly when it enters into the four walls and deduct it when it leaves those four walls (this latter is called back-flushing). Of course, you have to account for scraps or rejects and from time to time take a physical inventory within the four walls to reconcile any discrepancy, but this system eliminates a lot of transactions and overhead costs.

When Israel was established as a state, the ruler of Siam gave a gift of an elephant to the Prime Minister, but the Prime Minister of Israel sent it back with the polite excuse that his mother advised him, "Do not except a gift that eats." That is sage advice for controlling overhead. Do not allow programs to get started that require maintenance without contributing to your business. A small but telling example is planning social functions for employees. Instead, create an employee group to does that on its own time. If the group does it, well great; if it doesn't, your business will not suffer. Don't allow management to get involved. This is just a small example, but the list of unnecessary well-intentioned programs is long, and it takes away from the focus of the business, which is to compete and to make money.

One way to reduce indirect costs is to reduce the number of transactions. Transactions cost money. These costs are hidden because they occur in different places—on the factory floor, in the computer room, in report writing and reading, in data entries, etc. Some of these transaction costs are carried out by direct labor and therefore they are hidden from sight. By analyzing the flow of paperwork and the number of data entries for transactions, you can get a picture of what is happening in these areas. Having this analysis and using industrial engineering principles of eliminating or simplifying, you can achieve significant progress in controlling overhead costs. Many manufacturing companies have adopted the Japanese Kanban system or something similar, which is a tremendous help for eliminating transactions and paperwork.

Another way of reducing overhead costs is to create stability in design. This is very elusive, but there are large hidden costs due to engineering changes. These costs are seldom measured and if they are, they are usually understated because the transaction costs involved are not considered. The reduction of engineering changes starts with a corporate culture not to release products until they are fully tested. One way to measure how effective your designs are is to count the number of engineering changes that occur during the first year after product introduction.

Yet another engineering solution to reducing overhead is to design or redesign products to have fewer parts.

Here is an example of how this worked for a company for which I consulted.

A product having 700 parts was redesigned to contain only 200 parts. A count of monthly transactions (four transactions were considered normal in a department) was as follows:

Ordering transactions = 700 × 4 = 2800
Receiving transactions = 700 × 4 = 2800
Materials transactions = 700 × 2 ×4 = 5600
Materials authorizations = 700 × 4 = 2800

Total transactions = 14,000/month

After the redesign to 200 parts and changes to other transactions, here are the results:

- The factory issued blanket orders and communicated with vendors on just-in-time deliveries.
- A simple receiving and inspection procedure was implemented, which sent the receiving slip directly to accounting. The company only had to send one check per vendor per month.
- Delivering parts directly to the floor eliminated storing and staging materials in a warehouse, which eliminated many transactions.
- A Kanban system on the production floor eliminated material movement tickets.

Transactions were reduced from 14,000/month to 2400/month—a reduction of 6:1.

The above example included automation of the internal system for data entries in such a manner that one transaction automatically triggers the next one and maybe the one after that. This not only reduces clerical labor, but also speeds up reaction time and reduces errors. The best systems are those that only need to record data one time and generate reports or actions automatically by "talking" to the next step in the flow of data. These integrated systems exist today and it is only a matter of designing them and getting the hardware in order to have an overhead-efficient factory.

Integrated systems do not apply to the factory floor only. Order entry, production control, and customer service are also overhead costs that can be reduced by integrated data systems.

Reports on production efficiency and job costing require a lot of data entry and are mostly useless. These monthly reports after the fact can be very detailed but that does not mean that they are useful. If they contain a lot of data that nobody can do anything about, and they are read in the office by management one month after it happened, then they are a waste of money and time. The joke goes that a new computer system was bought by a company and an excited IT guy rushed into the president's office with the news that with this new program they could generate many more reports. "Oh, no," exclaimed the president, "Now we have to hire all those people to read the reports."

That is not to say that reports about measuring production or efficiency are not necessary. The message is that they should measure overall performance against budgets and only that. The reports should measure or highlight only variances that are out of "control limits." If there are large variances, they should be investigated on a case-by-case basis, rather than generating a lot of data about everything that is happening. That is true of the 80/20 or rather 90/10 rule of management by exception. You should only be looking at large variances from the norm. Measuring efficiencies should be done on the factory floor by the operator or supervisor, or by production control on a daily basis preferably with visual aids.

Manufacturing engineering should be making an initial investment in setting up operators charts on the floor that their lead or supervisor could maintain. Better still, have the operator maintain it. After this initial investment, these visual aids to measure output should be self-administered. This serves several purposes: no maintenance, immediate feedback, and visual representation. It is also a great incentive to operators to take an interest in their work. The charts can be statistical process control charts, output productivity charts, or number of rejects to measure quality.

11

Wage and Salary Administration

It is no secret that an important element of success in a business is the productivity of people in the organization. Therefore, it is important to compensate people according to the marketplace and give them the right incentives to make them more productive. No matter how hard management works, they cannot succeed unless they increase their leverage by having people that report to them work efficiently as well as they do. Management must have the proper tools to ensure that people in every part of the organization have the right pay and the right motivation to be productive. Only then can they expect them to be effective.

Naturally, it cannot be assured that everyone at all times gives 100% effort and is pulling in the right direction in a company employing thousands of people. (It is hard to do that even with two.) However, the closer you get to the goal of the right pay, the right incentive, and the right direction, the more effective and profitable the business will be.

In this chapter, we will talk about how to achieve the goal of correctly paying people in a manufacturing company. In Chapter 13, we will talk about how to motivate people after ensuring that they are fairly compensated.

The company must have a strategy of where it wants to position itself in the wage and salary marketplace. This positioning should be different for different skills. For example, a company who relies mostly on excellence in engineering for its profit may want to position itself in the top 75 percentile bracket for engineering salaries, and in the middle of the range for all others. A company that has problems with turnover of hourly workers may want to pay them in the top 60 percentile of the local market. No company works in isolation. There is a geographically different market for wages and salaries for different categories of employees. I emphasize geography because that may make a difference between benchmarking hourly employees and clerical or professional employees.

For example, in a small town a medium-sized company with low technology may want to emphasize the quality of life there and not compete dollar for dollar with IBM in an adjacent large city.

Wage and salary administration should consider all these when establishing the benchmark for each wage and salary category.

With that in mind, you should always pay at least as much as the average going rate and never pay less; otherwise, you end up with a below average workforce.

The next step in wage and salary administration is to grade each job and match it to the desired market level. It is essential to get surveys in order to determine where the job market is for every category. The surveys are by job description; thus, it becomes necessary to write a job description for each employee. Be careful in not getting too specific or spend needless time and effort in defining the job. In today's dynamic and fast-changing environment, it is counterproductive to have rigid job descriptions. Because job descriptions are mainly used for determining wages and salaries, it is best to use generic job descriptions similar to the ones in the surveys and try to fit each employee into one of these generic job descriptions. If that is impractical for a certain employee or group of employees, then a specific job description should be written, but keep in mind that in that case the surveys do not always give you a good guide as to where the "market" is for that category.

The great advantage of using generic job descriptions that mimic the surveys is that it saves a lot of time because these job descriptions are usually brief. There is no need to describe specific tasks to be carried out for each employee because these tasks may change and you want employees to be flexible and do the work where it needs to be done, rather than stick to specific tasks. Don't get enamored by bureaucracy and write a lot of specific job descriptions that nobody needs or uses. Specific jobs that cannot be found in the survey should be graded in between two most fitting ones, based upon skill and educational requirements.

When grading and specifying jobs, it is more important to remove "dis-satisfiers" rather than to stick to clichés. It is far better to use grades 1, 2, and 3, rather than the terms junior, intermediate, and senior. Nobody wants to be a junior clerk and stay there for more than 6 months, but Clerk 1 may be more acceptable. Stay away from being too specific. Don't use titles for job specifications like invoice clerk, accounting clerk, billing clerk, or cost accounting analyst.

A junior analyst may be a Clerk 1, an experienced analyst a Clerk 2, a senior analyst a Clerk 3, etc.

When job descriptions are used for hiring new employees, in addition to the generic description some key criteria must be added. That is a different situation and there is no need to write detailed job specifics for each employee just in case a position opens up for hiring.

All positions should be graded and a matrix of these grades should be created. This matrix should show all job descriptions and their grades so that they can be compared to each other across departmental lines. For instance, a grade 6 job could be that of an Accountant 3, Engineer 2, and maintenance manager having equal pay scales to be classified as grade 6. This matrix then should be published to allow transparency in the wage and salary administration. Specific ranges for each grade should not be published—that is too much transparency and more subject to changes.

It is important to make people understand that these grade matrices are not created as a matter of justice, but rather to acknowledge the market forces in your area. A grape picker will work harder than a maintenance worker will, yet the maintenance worker will be paid more. A plumber may not have the education of a professor, yet their take home pay may be the same.

Having established a fair and equitable basis and suitable grades and ranges, the human resources department is now tasked with administering the program. This consists of two parts. First, there is an annual review based on benchmarking surveys for each job category.

This is easy if you have done it before and only look for changes from year-to-year. It only becomes a problem when no basis exists, or the wrong principles were used, and the human resources department has to implement a new system because the old one was wrong or did not have enough data to be satisfactory. In that case, major adjustments have to be made to fit people into the correct grade and nobody likes major adjustments. It is up to Human Resources to adjust the scale over the years until it becomes fair and equitable based on the above principles. Some pay scales have to be redlined or grandfathered to avoid major dissatisfaction. Downward adjustments should never be made because that creates such bad psychology that the company would be better off to keep the employee at an artificially high level.

Upward adjustments should be made gradually to fit into the budget as shown next.

Having completed the survey and slotting everyone within a range, the Human Resources Department must administer the annual wage and salary increases. People want to feel good and make progress, even if it is only a little progress. That is why some inflation is good and that is why wages and salaries are going up every year, even if an individual is performing the same tasks as the year before. There could be some justification for employees being more valuable with more time on the same job, but that is not the only reason for annual increases in many situations.

Some companies adopt an annual review on the anniversary of when the employee started to work for the company. Others try to have the annual reviews always at the same time for every employee. The advantage of anniversary reviews is that the supervisor does not have to deal with too many at the same time. The advantage of having annual reviews at the same time for each employee is several and I favor that approach. (In that case, for new employees the raise in their first year should be prorated.) Here are the advantages of dealing with the wage and salary reviews at the same time each year for every employee:

- Wage and salary surveys are required for pay adjustments and they are more current when conducted before the fixed review time.
- Supervisors should rank their subordinates by performance during the previous 12 months and that can best be done when they all have their reviews at the same time.
- There can be no confusion or delay when wages and salaries are adjusted for each employee. Supervisors don't need to be reminded of anniversary dates, and there can be no disappointments when an employee for some reason does not get his/her review or raise on time due to lack of time or other "snafus" that happen in organizations.
- The amount of raises given should be governed by guidelines from the top and by budgets of what is available. This often forces a grading system by the supervisor to reward the best performers with more, but still have some budget left for others.
- The argument that it requires too much time for a supervisor to spend on reviews if they are all done at the same time only applies if the span of control is too wide (which is wrong) or if the guidelines for reviews are too cumbersome (which is also wrong).

It is necessary to review each employee's performance annually. Some companies want to separate this review from the wage and salary adjustment process in order to talk about career counseling and other issues of performance without bringing the money aspect into the conversation. This is a lofty idea, but very impractical. The most important people are those who are doing an outstanding job. You don't want to sit with them and praise them and then tell them to wait for their next raise. Conversely, there are people with issues that need attention and do not deserve a raise until those issues are resolved. That also is best addressed together with the annual wage review. Finally, give the poor supervisors a break from administrative duties and have them deal with all performance and salary issues in one sitting. There can be no doubt that the two are connected. If you did not pay your employees on Fridays, how many would show up to work the next Monday?

Most companies give annual raises to their employees. This is important in order to keep up with inflation and to reward people (it is assumed that most employees' productivity increases with time on the job), and also to adjust the marketplace.

The business plan should make an estimate of what the budget is for annual raises and then the human resources department should administer the process by giving clear guidelines to each supervisor on how to allocate the budgeted amounts. Before the increases are given to employees, the supervisors must submit the proposals to human resources to ensure that the guidelines are met.

Since human resources must administer and review all performance reviews as well as the increases, it makes sense to give all raises and performance reviews at the same time. Otherwise, the administration of this process can become too cumbersome and time consuming.

After grading and market values have been properly established and given to supervisors, they must now determine what wage or salary raise should be given to each employee reporting to them. It is desirable to give larger raises to people near the minimum of their range than to the ones near the top of their range. If that is not done, people near the minimum would never catch up and people near the maximum would max out. At the same time, those who performed very well during the year should get a bigger raise than the ones who did not. Complicating the issue is that of the budget, which will always limit the amount of raises that are available to each supervisor to divide between their employees. As you can see, it is not an easy task to determine each increase, and therefore the guidelines should reflect average increases with a range to administer them in a way to address inflation and to achieve budgetary goals for the company.

Of course, the percentage and ranges will vary from company to company and from year to year, but here is an example of guidelines for annual wage and salary increases. In this example, the budgeted annual average raise is 4%. The supervisor may start the calculation based on the following table and then make adjustments after estimating the raise for each employee and comparing the total with the total departmental budget.

Performance	Low Range	Middle Range	High Range
Outstanding	8–10%	5–7%	2–4%
Good	4–7%	3–4%	1–2%
Needs improvement	1–3%	1–2%	0–1%

Note that there are only three "buckets" for judging performance.

Many evaluation forms grade performance on a scale of 1 to 5 or 1 to 10 and include all kinds of categories to be filled out with these numerical ratings. Managers find that very difficult to administer and explain. It serves no useful purpose. In many categories, the employee is performing a good job, and cannot do outstanding work. For example, what can a payroll clerk do to achieve a rating of 10 on a scale of 1 to 10? Make out more paychecks? Most jobs in a company require a certain routine to be performed well, and that routine cannot be rated other than good or bad. Therefore, the best system is to rate performance in three buckets—outstanding, good, or needs improvement. Then the employee cannot be dissatisfied with a rating of 6 out of 10 or 4 out of 5, and quibbling over judgment calls between fine lines that do not really matter to the satisfactory operation of the department or the company. (More on grading in Chapter 12.)

"Needs improvement" refers to employees who are struggling with some (not all) aspects of their jobs. If the employee is not performing satisfactorily, he/she should not get a raise at all, but if he/she needs just some improvement, it is good to give a token raise.

In addition to the above guidelines, it is important not to allow supervisors the latitude to inject their personal preferences into the equation. Some supervisors want to be liked and would prefer to give large raises to everyone, while others are very tough on money due to their frugal nature. To take these biases out of the equation, the guidelines should specify that only a certain percentage of employees can be classified as outstanding performers (say, maximum 10% in any group) and needs improvement should not exceed 10% or there is a problem in that department, which needs to be addressed quickly.

Here are some guidelines on percentages to be allocated (this can vary from company to company):

Outstanding performers should equal no more than 5 to 10%.
Good performance should equal approximately 80% if nothing else than by default.
Needs improvement should equal less than 10%.

The evaluation of performance shown in this chapter should form the basis of the merit part of the equation for the salary increase. Using the three bucket system (outstanding, good, and needs improvement) will be one of the factors for distribution of the budgeted salary increases.

The other factor will be where the employee fits into the range of his/her grade. Special adjustments can be made for employees below the bottom of their range or to those who are topped out or above their range.

Each supervisor should submit a proposed budgeted increase before the annual review and have it approved before giving the increases to the employees.

This system does not allow for sufficient rewards to outstanding employees because of budget restrictions. There should be a different way to reward truly outstanding performers of the month or year with one-time bonuses or rewards (see Chapter 13).

12

Performance Reviews

The performance review should be given at the same time as the wage or salary increase. Annual performance reviews are part of corporate culture and properly handled serve a useful purpose. Unfortunately, most people receiving them or giving them hate them. That is because most performance review formats try to accomplish too much. Designing the form of the performance review, people over time generally incorporate any ideas that could be useful "just in case they apply." Thus, most performance reviews contain narrow definitions and checklists to be filled out with grading each parameter to force the reviewer into making judgment calls about subjects that may apply to only a handful of people.

It is widely acknowledged that performance reviews should not contain any surprises. Why is it then that most forms and discussions have to be long and cover so many issues? Why do almost all performance reviews insist that the employee review himself? The employees know that their performance reviews will affect their raises. There is no way employees will own up to their own shortcomings. That will have to come from the boss.

Many performance reviews view the process as a two-way street where employee and boss sit down together to exchange their views, but that is a false assumption. The real issue is that the employees' performance is being reviewed, not the supervisors' or the company's.

Most performance review procedures include feedback from the employee. That makes the whole process complicated and awkward. If the employee has some issues or dissatisfaction with the way he/she is treated, there should be other forums to air that than the annual review when the employee gets a raise. How can an employee then make his/her voice heard, ask for a raise, a promotion, or complain about something?

The performance review should not be the only avenue for that. An employee should not have to wait a year to voice a request or dissatisfaction.

That is what "listening sessions" during the year should accomplish. Hourly employees should have monthly meetings with their supervisor on the factory floor, and salaried employees should have a type of MBO program where they meet formally at least once every three months with their supervisor.

The performance review is the worst time to act as a "listening session." It dilutes the purpose, which is reviewing performance during the year and plans for the future.

I suggest that there should not be a multitude of questions about how the employee feels about every aspect of his/her job. One general question asking about feedback from the employee about any areas of concern, but narrowly defined as having to do with job performance, should suffice.

Managers are just as frustrated with the performance review process as employees are if the process is lengthy, cumbersome, and trying to do everything for everybody with a multitude of questions that must be filled out on complex forms that sink under their own weight.

Since there should be no surprises during the performance review, the format should be simple. The employee's performance should be rated, future plans should be discussed, any necessary improvements should be noted, and the employee should be asked for any feedback regarding his/her performance and future plans. At the end of the review, the wage or salary adjustment (usually an increase) should be given to the employee. A simple form should suffice as follows:

- Rating of performance during the year.
- Discussion of performance highlights or deficiencies.
- Plans of duties for next year (could be blank space to be filled out).
- Any educational or training needs to be discussed.
- Feedback from employee (blank space to be filled out if necessary).

No specifics or checklists are needed on the form because each case is different. Very little preparation is needed if the employee and supervisor have worked closely together in an open relationship during the year. If that is not the case, no amount of checklists of attributes will help the relationship.

Rating the performance should not be done on a bell curve in a department, where the supervisor is forced to rate some employees on a lower scale due to restrictions of the salary budget. If in a department the average increase cannot exceed 4%, it does not follow that the supervisor should rate some employees below average and some above to arrive at a budgeted raise. The performance rating should not be affected by the salary increase given. If all employees in a department perform well, they should all be rated as good or outstanding. (No more than 10% can be outstanding by definition.) It is true that due to budget constraints and grade range considerations, the wage and salary increase is only slightly affected by performance; however, a performance rating of outstanding should enhance the employee's chances for advancement or a one-time bonus.

Details of what part of the job was performed well or not so well are left to a blank page discussing highlights and deficiencies of the performance during the year.

The most unsavory systems encourage the supervisor to make notes of mistakes or problems during the year and then hit the employee over the head with these complaints during the performance review. The same principle applies to praise. Don't wait for it, give it freely when it is deserved.

There have been many different fads and social experiments tried to enhance the performance review. One of them was called 360 degrees, where peers evaluate the performance of the employee. This just leads to bad feelings between peers. Another system was for the employee to do his/her own review. Sometimes this is combined with comparing the self-review with that of the supervisor. None of these experimental systems had any traction. Every few years another system gets a trial. What needs to be understood is that no format can substitute communication between employee and supervisor, and the performance review is not a ritual. It should be a review by the supervisor (representing the company) of the performance of the employee. To have the employee write his/her own review does not make sense.

Before embarking on complicated systems for performance reviews, let us review the purpose of doing them and what the company is trying to achieve by the process.

Is it necessary to have a formal meeting at least once a year, when the communication between employee and supervisor should be an ongoing process during the year? If there are no surprises, why continue the practice?

The answer is that if the reviews were eliminated while every other company has them, it would send the wrong message to employees. The second reason is that it should be connected to salary increases or adjustments, and almost every company does that annually. The third reason is that it is a good time to review educational needs and ambitions.

Some theorists on management philosophies disagree on the subject of giving salary increases at the same time as reviews, arguing that it would distort the dialogue between employee and supervisor. These same people then burden supervisors with separate reviews for salaries and performance and then devise complicated systems to try to force the performance review to be an annual ritual of predetermined questions. They do not trust the supervisors with covering every aspect of performance and every subject that can arise between employee and supervisor. If the employee and supervisor cannot be trusted to fill out blank spaces with the most important criteria, no amount of force-feeding many questions will improve that dialogue.

Another reason for performance reviews is to warn employees of impending termination for bad performance and thus protect the company from lawsuits. That is a valid concern, but I contend that the process of weeding out or reassigning unsatisfactory employees to another task cannot wait for an annual review process and should be part of personnel policies that deal with that question as soon as it starts to become a problem. Don't wait for the annual review.

Finally, we get to the crux of the matter of communications between employee and supervisor and the importance of training and close supervision. Each supervisor should have the right tools to communicate with and motivate each employee. These tools are training sessions, listening sessions, MBOs, etc. that occur regularly during the year, and should not wait for the annual ritual of a performance review.

Following is a suggested performance review form, consisting of only one page and devoid of any specific questions or grading. Because every situation is different, it is best to leave many blank spaces to be filled out only if needed and with the specifics that matter for an annual review of performance.

PERFORMANCE REVIEW FORM

Name of employee: _____ Date: _____

Major accomplishments during year:

Overall rating:

Outstanding　　　　　　　　　Good　　　　　　　　　Needs Improvement

Plans for next year:

See MBO if applies. Include specifics if employee's performance "Needs Improvement."

Career plans and/or educational plans for next year:

Employee comments regarding performance:

Copies: Supervisor, Employee, Human Resources

13

Motivation and Productivity

One of the most important elements for success of a business is the productivity of employees. No matter how hard management works, it will be useless unless they use leverage by improving the productivity of others in their organization.

It is not only important for people in the organization to work hard, but it is also important that they work on the right things and in the right direction; that is what I mean by productivity. It means efficiency focused in the right direction. You can be the fastest runner, but if you do not run in the right direction, you will not get to your destination faster.

It is management's task to provide the right direction for productivity but that is not enough. Management must also provide the tools and processes to motivate people to put forth their best efforts.

There is a difference between motivating hourly employees and salaried employees.

Most organizations classify hourly employees as those who manually work on the products, and I will use that definition here.

Hourly employees who perform manual labor or repetitive tasks have their jobs defined in very narrow terms and are not expected—nor is it desirable for them—to deviate from the given tasks. Therefore, their motivation has to be tailored to how well they perform the given task. In contrast, salaried employees in many cases are not confined to routine work and can be expected to use their skills in innovative ways. Their contribution—or lack thereof—can have a larger impact on the bottom line. This does not mean that hourly employees are not important to an organization, but their contribution is measured differently and their motivation should be addressed differently in most cases.

Before addressing hourly worker productivity, I would like to refer to two experiments on this subject.

Robert Owens, an English philanthropist 200 years ago, inherited a factory. He was not an industrialist, but he was a great student of human nature. He decided to increase productivity in an innovative way. He arrived at his new plant with three spools of colored ribbon. One morning the workers came to work and found a ribbon tied to each machine. Some machines had one color ribbon; some had another color. Everyone buzzed with curiosity. After several days, the word got out. Red ribbons were tied to all machines that were producing above the factory average. Green ribbons represented the average output. Yellow ribbons meant below average. Owens made or implied no threats. He was merely using the ribbons to let people know how they were doing. Within two months, every ribbon in the factory was red. Production had never been so high, and morale had never been that good. The colored ribbons enabled workers to keep score and to compete with each other.

Another famous experiment that you may have heard of was the Hawthorne experiment, at a General Electric factory between 1924 and 1927. The objective was to find out what variable factors affect productivity. The operators were informed about the experiment and asked to participate and give feedback. There is a lot more about this experiment that the reader could find by researching the subject, but for the purposes of this book, I will only detail the result.

The operators were put on piecework for eight weeks.

Output went up.

Two 5-minute rest periods were introduced.

Output went up.

The rest periods were increased to 10 minutes.

Output went up.

Six 5-minute rest periods were added.

The operators complained that their rhythm was broken. Output fell slightly.

Return to rest periods for 10 minutes was reintroduced and company supplied hot meals.

Output remained the same.

Various experiments with lighting and working conditions were conducted. Output kept going up regardless of lighting changes and working condition changes.

Finally, all improvements were taken away and the operators went back to the old physical conditions at the beginning of the experiment: work on

Saturdays, 48-hour workweek, no rest periods, no piecework, and no free meal. That state of affairs was measured for 12 weeks. (Such working conditions were typical during that time.) Output was the highest ever recorded.

The findings were that there was a general upward trend in productivity completely independent of any changes made. This was due to the group in the experiment being isolated and observed. Employees in the group knew that they were being measured and that somebody cared. The group formed a social atmosphere that included the observer who tracked their productivity.

The lesson learned from these two examples is that people do not work for money alone. There are additional motivators like personal recognition, sense of achievement, and social factors like a sense of belonging. Working conditions make no difference unless they are unacceptable. To motivate people you must pay them attention, measure and give feedback on their performance, and recognize their efforts.

Measurement and feedback systems should be tailored to each job and must strike a balance between administrative costs and effectiveness of the plan. One system does not fit all. In some cases, piecework is best suited for motivation, in others, an assembly line is most effective, and in many situations, a simple visual chart that can even be self-administered can be very effective.

A word of caution about piecework—because it involves money, it must be more precise than any other system. It becomes more difficult to establish time standards and to measure output. Arguments arise regarding a full day's work measured by time standards of every activity, treatment of rejects (sometimes quality suffers), accuracy of time standards, etc. Compiling reports and calculating pay have administrative costs that can be significant. Therefore, only consider piecework if the work is highly repetitive, can be easily measured, does not have many variables, is performed all day, and has been proven effective in that industry. If applicable, piecework is the best way to increase and maintain high levels of productivity, but good applications are not easily found.

Sometimes a simple chart in front of the operator showing daily output that is marked by the operator is sufficient to establish baselines and goals.

Any system of measurement falls apart and loses its effectiveness if nobody is watching it and if there is no feedback. If supervision does not care, why should the operator?

The closer the intervals of feedback, the more effective it is. Theoretically, the best result would be obtained if a lathe operator would receive a coin after each piece of machined part drops into a bucket. An example of the importance of frequent feedback follows: If a painter is given 5 days to paint a house, he may not finish it in time, but if a section is marked off for each day, the chances are much better that the job will get done in time.

The key principle of establishing standards of performance is "attainable goals," which bears repeating in this chapter. An attainable goal (whether for time standards for hourly workers or a management objective for a scientist) is one that can be achieved by constant steady effort without letting up during the time allotted. It is not easy to achieve, but it can be beaten by super effort. If a goal cannot be beaten, it is not a reasonable motivator.

It is not always easy to establish standards, but you must try to do it for as many jobs as possible.

It is easier to establish standards for repetitive jobs, but we cannot give up on doing it for non-repetitive ones.

Here are some examples on how to establish standards for variable output:

A call center supervisor wanted to measure the number of calls handled by each operator. Some calls were longer than others and it was difficult to establish a standard. There were 15 operators handling calls and the supervisor decided to put up a chart showing the number of calls handled daily by each operator without establishing a standard time for each call. Each operator recorded the number of calls per day. After a week, the numbers averaged out and it became clear which operator handled the most calls and which one the least. Thus, competition was created and output went up for the group. The supervisor was able to reward the best performer of the month and to find out the problems with the lowest performer in the group. If the lowest performer consistently stayed at the bottom, the supervisor knew that he/she was a misfit for the job and took corrective action. Most of the system was self-administered by the operators after the supervisor created the charts.

A research engineering supervisor found it difficult to establish performance objectives. Each scientist was working on different projects and the length of time to complete them varied for each project. The supervisor tried to find a common thread and realized that every job had one thing in common and that was the final report. The supervisor also knew that reports were the most important output and was the thing most scientists were loath to write. The supervisor devised a chart showing each scientist's name and beneath it the name and number of completed final reports for the year. The chart started in January and went for 12 months. Reports increased in volume and productivity increased as the year went by.

A supervisor for insurance clerks was plagued by errors made in calculating claims. The supervisor talked to the clerks about it, but the error rate did not change. The supervisor then made a monthly chart with the names of each clerk and showed the number of errors during the month. The idea worked and the number of errors decreased dramatically.

Following are the steps necessary to introduce systems of motivation into a group.

- Sell and explain the system to the group.
- Apply controls to the smallest grouping as possible.
- Provide visual comparisons.
- System must be easily maintained.
- Self-administering is very desirable.
- Constant attention and feedback are essential.
- Positive recognition is far better than negative feedback.

While all of this seems like a lot of work and effort, the resulting increase in productivity by far outweighs the cost.

Motivating salaried employees needs different approaches than motivating hourly employees. Of course, sometimes there is an overlap because the definition of hourly and salaried varies from company to company. In this context, the term "hourly" applies to employees who do mostly repetitive or manual tasks, and the term salaried refers to employees with a broader level of decision-making, requiring initiative with less rigidly defined tasks.

Nevertheless, the same criteria for motivation apply, only the measurements are different. Every employee needs to be motivated by measuring his/her performance, feeding back the results, and paying close attention to rewarding good performance. There are many systems for measuring performance of salaried employees. The worst is the annual performance review, which is too little too late. Feedback should be much more frequent. The best system I found and use myself is Management by Objectives (MBO).

MBO is a system of setting objectives for salaried employees. It could apply to all salaried employees because by definition, they all have a supervisor who should set goals for them and these goals should be reviewed often during the year. Normal frequency for MBO programs is every three months. The program is designed to ensure that salaried employees are aware of their priorities and that these priorities agree with the business plan for the company. Since it is set up and reviewed every three months, it is designed to ensure that the plan is on track and on schedule and to identify and correct any shortcomings during the year of meeting the business plan. The MBO program is also a motivational tool for salaried employees to provide the measurement and feedback that every employee needs and to make sure that supervisors are aware of how salaried employees are performing and how they feel about their objectives. It is important for an MBO program to concentrate on the most important objectives. Therefore, only top priority objectives should be included and the format should be restricted to no more than 3 to 5 objectives every three months.

Some of the objectives will be numerical, some strategic. Some priorities of the business cannot be completed in three months' time. In that case, they should be broken into elements that can be completed during the three-month period. For instance, a company may decide to build a warehouse in another region. The objective for the next three months for the VP Operations may be to identify the location, line up the contractors, and submit a plan. The building of the warehouse—if the plan is approved—will be the subject of consequent MBOs.

An MBO form should be filled out every three months for salaried employees by their supervisors. This form should include a listing of the objectives, the time of completion, and a rating against each objective. During a meeting with the employee, the supervisor discusses the previous three-month MBO and presents the next three-month objectives. Sometimes when the previous objectives are not fully completed, they carry over for the next three months. This discussion about objectives accomplished and new objectives set is really a mini performance review every three months of the top priorities for each employee. That helps to focus the employee and the supervisor on what is important in order to accomplish the business plan.

In addition, this formal meeting every three months should be a listening session as well, where the supervisor listens to the employee's thoughts and concerns on a one-on-one basis.

A simple rating system should be part of the MBO form consisting of three grades:

- Incomplete
- Satisfactory completion
- Exceeded expectations

Everybody wants to be—and should be—rewarded for outstanding contribution. If an employee exceeds expectation in key priorities, there should be a system designed to reward that employee as part of the MBO program, even if it is only a small token of recognition, like lunch with the supervisor.

Following is a typical MBO form:

MANAGEMENT BY OBJECTIVES

Year_____ Quarter_____

Name_____Department_____

No.	Objective	Completion O, S, or I	Remarks
1			
2			
3			
4			
5			

Overall rating of performance = OUTSTANDING, SATISFACTORY, OR INCOMPLETE

Supervisor_____ Date_____ Employee_____

Outstanding performance is not always possible given that objectives when completed are satisfactory most of the time and cannot be "overachieved."

Incomplete objectives may be carried over into next quarter, if still deemed a priority.

14

Benefits and Incentives

Fringe benefits fall into two categories—satisfiers and motivators. The satisfier part is easy to determine. Just like with wages and salaries, the company should survey the local market and design a benefit program that is average for that market. Notice that in case of wages I advocated to be slightly above average for that category, but for fringe benefits the average is sufficient. The reason for this is that benefits are satisfiers, not motivators. Once a benefit is given it can hardly be taken away; it is taken for granted. Therefore, above-average fringe benefits can be problematic when times are tight and they have to be reduced.

Wages and salaries should be at or above average and they are motivators because there is room for more when annual increases are due. There should be sufficient budget available for other motivators as well. If you give too much away in fringe benefits, you may not have enough room for motivators.

There has been a tendency for leading edge technology companies to give fringe benefits that are above the normal. These fads last a couple of years in areas where companies compete for talent and they include babysitters, free meals, and even swimming pools and exercise machines, which are not only costly, but are hard to administer. It is doubtful that these benefits play a big role in employee recruiting or retention, and I would recommend translating all of that into money and offering higher salaries and rewards for outstanding performances instead. For most companies it is wrong to offer above-average fringe benefits or activities (like free meals) that require administration and maintenance. Rather, concentrate on motivators tied to performance.

It is very important to reward outstanding employees because it motivates everyone else who would like the honor and the reward of being outstanding. The best way to do that is with one-time bonuses or recognitions.

There should be a budget for these programs because wage and salary increases give very little in rewarding excellent performance in a low inflation environment when general salary increases are low and there are other factors competing for raises, like time in grade.

The first principle of incentives or bonuses is that they should be one-time awards. This is what distinguishes them from benefits. While benefits are "satisfiers" and can rarely be changed or withdrawn, the advantage of one-time incentives is that they are not expected to be repeated.

For hourly employees, outstanding work recognition is not required for piecework because it is automatic, but as we discussed previously it seldom applies. If it does not apply, the company should design a system of rewards that a supervisor can easily administer; employee of the month is just one example. If work is measured against a standard and records are kept of performance (as they should be), then the best performer in each work center should get a bonus or reward every month. When several assembly lines compete with each other, the best line for the month should get a bonus for everyone on that line. Competition for rewards is very desirable for any incentive system and to reward the best assembly line is a great way to implement this.

Any bonus or incentive system should be easy to follow and should be transparent so that each individual or group knows where he/she stands at all times.

My favorite reward system for hourly employees is a decentralized approach that gives supervisors a budget and guidelines of a self-administered system that requires very little administrative work. The guidelines tell the supervisor how often and how many rewards he/she has to give and what the budget is. The supervisor can then choose between the following rewards: A catalog from which to choose gifts, time off work, dinner, or a trip for two. A photo posted and a handshake in front of the peer group are also excellent motivators and should be encouraged. In this system, there is a danger of favoritism by the supervisor and this is a drawback, but the simplicity of trusting the supervisor without any paperwork or administration should outweigh that disadvantage. There can be abuses to any system, and the best way to mitigate this is by education and occasional audits.

For salaried employees the MBO system is an excellent vehicle for awarding bonuses. If an MBO program is in place, then these bonuses can be given every three months to outstanding individuals who exceed the goals set in the MBO. Of course, the system should be designed in such a way that the same employees could not continue to get bonuses all the time; otherwise, it becomes a routine giveaway and loses its purpose.

For sales employees, a commission system is the best program because it ties performance to rewards the most direct way.

For key employees, stock options are the best motivators. One word of advice about stock options—if awarded they should only be long term, for the obvious reason; the company does not want employees to act for short-term objectives, nor do they want employees to cash in stock options and then leave.

15
Role of Engineering

The role of engineering in a manufacturing organization is both to design new products and to work on engineering changes necessitated by manufacturing or customer needs. These needs could be marketing requests, quality problems, manufacturing problems, vendor problems, or tasks to reduce costs. It is not easy to prioritize all these support needs and still have time to design new products. If the department is large enough, my recommendation is to divide the department between new product design and sustaining engineering, with each having separate budgets. After that, the new product design function will prioritize which new design to work on, and the sustaining engineering function will prioritize which needs to fulfill first. Engineers and support personnel would be permanently assigned to each function, with some overlap in specialized fields.

For both new designs and redesigns, design for manufacturability is most important. This does not come naturally to design engineers, nor are they always taught this subject in engineering schools. Therefore, the company must have periodic training programs for engineers to learn design for manufacturability. In the modern factory using robots, there is a whole science of how to design for robotic assembly as well as for lowest costs. There are training programs available that should be investigated and one of them selected because there is little chance that a company has these training skills available in house.

At a minimum, design engineers should be given a book on design for manufacturability as required reading and their supervisor should lead discussions on the required reading book.

It is also a good idea to expose design engineers to the production processes that are used in the company.

It is important to note that most costs are assigned and most of the quality determined when the product is designed. Once this is

understood, it is easy to see why design for manufacturability is one of the keys to being competitive.

Techniques for designing for manufacturability include:

- Consult with manufacturing engineering and with service at early stages and throughout the design.
- Minimize total parts count.
- Use common components to other designs. Commit to a standard parts program.
- Consider testability.
- Use a modular design where the module can be tested by itself.
- To aid robotics, use stackable design principles from the base up.
- Establish parts positioning with guides, pins, or recesses.
- Avoid multi-motion insertion.
- Provide chamfers or tapers for positioning.
- Design for lowest cost fastening. Avoid screws where possible.
- Use common size and types of fasteners.
- Avoid labels where possible (if not possible, combine labels).
- Exaggerate asymmetrical parts design.
- Consider serviceability.
- **Specify largest allowable tolerances.**

Note that this is only a partial list and not in priority order.

An important responsibility of engineering is to test the design before releasing it to production. It is estimated that if a design problem occurs during production, it is ten times more costly to fix it when it is caught before engineering release. If the problem is found in the field, it is 100 times more costly to fix it, and if it causes litigation or product recall that cost could run into the millions of dollars.

Engineering must be involved in new product introduction. This function is usually a project that is run by a task force. The product manager, with participation from manufacturing engineering, should chair the task force. Design engineering should be responsible for running a pilot program to ensure new product integrity and manufacturability. After the pilot program is successful, design engineering should formally release it to production. After that, sustaining engineering takes over the maintenance of engineering documents and engineering changes.

16

Marketing and Sales

The difference between marketing and sales is that marketing goals are long-range and strategic, while sales goals are short-range and tactical. That is a huge difference. It obviously requires different personalities and skill sets. Yet, very often people migrate from one to the other or are promoted from sales into marketing or the other way around. These can work only if the salesperson is capable of long-range planning, or the marketing person is capable of selling. Marketing people are the planners; salespeople are the implementers. Marketing is a staff function while sales is a line function. Some staff people are capable of assuming line responsibilities and some line people are good at staff functions, but in the selection process, the personalities and skills required should be kept in mind before making them move into unfamiliar roles.

Marketing people should have different incentives than salespeople to motivate them toward the correct goals. These incentives should focus on long-term product objectives rather than short-term sales objectives.

The proper organization in a manufacturing company should be that marketing operates through product managers segmented by product lines. Sales should be organized by regions to be close to the customers.

The marketing organization is responsible for product management, sales and distribution strategies, and sales support. The marketing department is the visionary in the company and as such, must play a key role in formulating the business plan in conjunction with the CEO. Marketing must have an overview of the market forces at work, competitive pressures, customer perception, and the company's strength and weaknesses versus competition. From this overview, marketing must come up with the winning strategies to formulate the business plan.

The purpose of the business plan from a marketing perspective is to increase sales and maximize profits. It must take a long-term strategic view and shed unprofitable products while investing in and developing new ones for the future. Marketing must keep in mind that the objective of a corporation is to make a profit and not to increase sales by sacrificing profits (unless the short-term strategy is to increase market share regardless of profit margins). **This motive—of profits being more important than sales—is where the sales and marketing approach is completely different.**

To a good salesperson, closing the sale at all costs is of paramount importance. If the salesperson would stop and think whether a sale should be made with minimal or no profit margins, he/she would not be a good salesperson and should look for another career. A good salesperson will try every avenue to secure the sale. In order to foster this attitude, his/her compensation and incentives are also structured purely on sales revenues, rather than profits. That sounds wrong, but it is correct. Trying to argue overhead costs and strategies over each sale is futile. It is best to let the salespeople know the rules and let them operate within those rules as best they can. The rules are that sales people should never be allowed to set pricing or quote prices without authorization from marketing. Marketing is the strategic function in charge of pricing and profits.

Marketing must determine the price for every product, based upon the strategies outlined in the business plan. Market share strategies call for lower prices, while strategies for maximizing profits need higher prices. Marketing should set the strategies in each case for each product.

Considerations about product life cycle should affect pricing as well. Sometimes lower prices mean higher volumes, while higher prices most often result in lower volumes, but higher profits. I am only pointing this out because it is wrong to take the cost of products and apply the same formula or add on the same percentage profit to each product category or each product. While convenient, it is also wrong to do that. Pricing decisions are strategic.

The General Manager can overrule pricing decisions when business conditions change. If there is no capacity to manufacture more products, the General Manager may opt to increase prices or be very selective in taking on new business. On the other hand, during lean times instead of laying off people it may be necessary to lower prices, even below standard cost, to cover the overhead for a short period of time.

Marketing must assume the role of overseeing the implementation of the business plan. That is where product managers come into the picture.

Product management is a matrix function that is responsible for a segment of the market. This is an essential function in order to make sure that sufficient attention and focus is brought to a given segment. Product managers are constantly fighting to get enough priorities for their product lines, from engineering, sales, and manufacturing. "Fighting" sounds like a bad idea within a corporation, yet in any matrix function there is competition for resources, and that is what is meant here. Competition for resources is unavoidable in a matrix organization. It is healthy for the corporation because it is a way to prioritize resources.

Product managers chair new product introduction meetings for their market segment and are responsible for marketing strategies and budgets for introducing new products, and maintaining their existing product presence in the field. Product managers should have responsibilities of profit and loss of their product segment. They need the help of finance to ensure proper cost allocations for their product. (More details on cost allocation can be found in Chapter 17.)

Product managers are also responsible for managing every product through its life cycle including the decline and phasing out of a product. It is not a thankful task.

Sales activities provide the revenue and lifeblood of the business. The whole organization must be sales and service oriented because without sales there is no company. The best way to organize sales is to get salespeople as close as possible to the customers and let them focus on the customers rather than having administrative duties. While your salespersons are filling out internal reports, they are not selling. **Make sure that salespersons sell, sales managers manage salespeople, and support staff support all functions that are not directly related to selling.**

If there are enough salespeople in a region that justifies a sales manager, then it is important to find someone with managerial skills who understands that the advantage in making 10 salespeople more productive is worth more than him/her making some sales. The compensation for the sales managers should reflect this philosophy. Under no circumstances should the sales manager be rewarded commission for making a sale. He/she should get a percentage of the commissions that the salespersons earn who report to him/her. That way there is no conflict. Otherwise, the sales manager may keep the juiciest accounts and not pay much attention to what his/her subordinates are doing.

Very often, your best salesperson is earning high commissions but just because he/she is such a good salesperson, he/she would not necessarily become a very good sales manager. He/she should only be promoted to sales manager if he/she has managerial skills as well as sales skills.

Top salespersons not suited for management should be given a "parallel path" and it should be understood that sometimes the top salesperson would earn more than his/her manager would. Everybody wins!

It is important to understand that working on strength is better than trying to shore up weaknesses. This applies to every aspect of corporate life and is especially true of salespersons. There is an anecdote with a lot of truth to it to illustrate this point.

Someone poses a problem to a sales manager: You have two territories, North and South, and you have only two salespersons to cover them. North is a very good territory, while South is a marginal territory. Of the two salespersons, one is better than the other is. Which one do you send to the North and which one to the South territory?

This is a trick question. The correct answer is, send both of them to the North until that territory is saturated.

The lesson here is that you should always work on strength rather than shore up weakness. If you send your best salesperson to a weak territory, the results will not be stellar.

Of course, there can be cases of "missionary sales" where it is strategically deemed necessary to open up new sales territories, but this should be a strategic decision to sacrifice short-term gains for a longer-term goal.

Often in a given region, there is not enough management work for a sales manager. Ideally, the salespersons in that region should report to another region or directly to the head of sales or even the head of the marketing department. The span of control rules can be relaxed in the case of commission-compensated salespersons because they do not need much supervision if they work on commission. Partial sales managers, managers who sell as well as manage, are not very effective supervisors. Their compensation package (salary plus commission) also gets complicated.

Sometimes it is not possible to compensate salespersons with commission only. In the early stages, the salespersons may not have enough income until the territory is mature enough and enough time has been spent in the territory. The lead-time for getting accounts may be rather lengthy. The company may need "missionary" work done by the salesperson to open up new territories. Some companies adopt a combination package of salary plus commission. There is no best way, but the objective should be to keep increasing compensation by commission as fast as possible, and to keep the scheme simple, transparent, easy to measure, and with a minimum of administration.

Sales organizations are unique in one respect. Most salespersons work on a commission tied to sales. There are many schemes where there is a combination of salary plus commission, but commission usually plays a major part in the salespersons' income. If we reward salespersons with incentives tied to sales volume only, it is difficult to get them to do anything else. If we load them down with reports, ask them to do surveys or market research, they will not do these activities well. Yet, they are the ones closest to the customers and the organization needs to gather these data. This dilemma can only be solved by understanding each unique situation and making the right choices on how to provide support for the salespersons to minimize the distraction for them and still get the right answers for the organization. That is why I believe that order entry and customer service should report to the sales organization because they can best support the salespersons. Order entry and customer service can form the basis for a sales administration function that provides administrative help to the salespersons, thus relieving them from administrative work.

Sales administrations should minimize administrative effort or research required from the salespersons wherever possible. If market research and forecasting data can be obtained without bothering the salespersons, then that is how the program should be structured.

There is one activity whose importance to the bottom line is underestimated by most corporations because the cost is hidden. That activity is accuracy of forecasting.

Mark Twain once said, "It is difficult to make predictions, especially about the future." While that is true, the closer forecasts get to actual sales, the better the results are in controlling manufacturing costs.

In most corporations, there is a monthly forecast meeting where it is the job of the sales manager to forecast sales for each product. The manager throws some numbers together the night before or the morning when it is due, and you are lucky if he/she even attends the meeting. The manager's priorities lie elsewhere, and he/she certainly will not ask the salespersons for numbers because salespersons are notoriously optimistic. (Surely, this does not happen in your organization.) The consequences of bad forecasting cascade throughout the manufacturing organization, from ordering materials, to upsetting current or future production schedules, to nondelivery of products. Sometimes a promotion for a certain product or product line is not taken into consideration and the factory runs out of the product. Sometimes after a promotion, the forecast is not adjusted downward and there is overproduction, which results in too much inventory and uneven production schedules. These problems are costly but they are hidden costs.

Recognizing that forecasting cannot be accurate, the question is how can it be closer to actual? The answer is with pre-planning and effort. The procedure to prepare a forecast should start a week or a few days before it is due, and there should be an effort made by the salespeople and their manager to get as close as possible to the actual number that will be sold. It is a thankless job, it happens every month, and it is time consuming. Efforts to reward or punish salespeople for the accuracy or inaccuracy of the forecast have been tried and failed. It is up to top management to educate sales supervision about the importance of this function, and to create a procedure to ensure some semblance of compliance. At a minimum, the sales manager or administrator should be required to attend the meeting and to show what effort went into the forecast. Sales must forecast each product by volume, not total dollars. When a sales manager said that his/her forecast was pretty close in overall dollars, a manufacturing manager replied, "We are not building dollars."

Part of company culture on teamwork is to make key executives understand the problems and the effect their actions have on their peers. Sales forecasting is one of the best examples of how neglecting a key activity has a negative effect on manufacturing and thus on customer service and profits.

17

Financial Controls

The job of finance is to set up controls to measure and report deviations from the business plan and to provide timely information on the state of the business both inside the company and to the outside world. The finance department must design financial controls and reporting systems to accomplish this goal.

It is important for financial controls to organize the company into manageable entities. From the standpoint of budgeting and financial control, the company should be organized into profit centers and these profit centers should be divided into cost centers. Profit centers should be market oriented. This becomes clear in large corporations as they organize themselves into business units based on the market for their products. Of course, in a small company serving the same market there is only one profit center.

The number of profit centers depends on the diversity of the market served, and on the span of control that the CEO wants to create.

Within a profit center, the further division is the cost center from a financial point of view. The reason for cost centers is to assign responsibility along with authority for budgeting and financial controls to the smallest entity, that entity being the cost center. Thus, the definition of a cost center is where a supervisor is responsible for the budgets of that entity.

The finance department sets up controls and a reporting system against budgets for each cost center, and thus maintains vigilance for achieving the goals set out in the business plan. By adding up the cost center financials, the finance department will then issue profit and loss pro-forma reports for the profit centers.

In a matrix organization (and they are all matrix organizations) there are indicators other than pure profit and cost centers that need to be measured. The finance department is in the best position to measure and give feedback about many of these indicators for adherence to the business plan. These key indicators may not all be strictly financial, but the finance department should be able to gather the information and act as the "watchdog" for the organization. An example of this is reports of profitability by product category or market segment, rather than cost center.

Budgeting should be bottom up by every cost center working up the details. After that is done and submitted, the budgets will be decided top down to allocate resources within what the company can afford to meet the business plan.

The chief financial officer (CFO) plays a key role in resource allocations and presents alternatives to the CEO. After finalizing the top-down budget, the finance department breaks down the totals into budgets for each profit center, and within each profit center to each cost center. The financial analysts work with each cost center supervisor to compare the top-down plan with the bottom-up plan, and then explain to them why they can't have all they want.

Cost accounting is a key function within the finance department. The company must know the cost of every product. The conventional costing systems break down the cost of material, labor, and overhead. Usually material and labor costs are straightforward, but still the cost accountant must verify the validity of the calculation. Overhead is more difficult to calculate because every part of every product may incur overhead costs in different ways.

Overhead is everything that is not counted as direct labor or material, and the finance department must make sure that every expense is accounted for when calculating the overhead. The simplest way for a profit center to allocate overhead would be to take all the expenses and allocate them to direct labor. Thus, every hour of direct labor would "carry" the same burden of overhead. Unfortunately, it is not that simple. Material is usually a major part of the cost, and it is logical that material should carry some of the burden as well as labor. Within a profit center, different operations and different types of machines may actually cost more in overhead than others. Why is it important for a company to know how to allocate overhead more accurately? The answer is that a company must know its costs in order to price its products competitively and profitably; otherwise, it will lose either business or profits. If one product carries more overhead costs (burden) than its fair share, that product may be priced out of the market, while others that carry less than their fair share may be sold below their costs. It should be obvious that wrong overhead allocations lead to wrong decisions in many aspects of the business.

In allocating overhead, the cost accountant must determine what causes the expense to be necessary. If the material content is much larger than the labor content, then obviously materials should carry more overhead than labor. Conversely, if the operation is labor intensive, then a larger proportion of the overhead should be allocated according to labor hours spent. Another factor occurs when large capital expenditure is needed for a special purpose machine. The operations carried out on that machine should carry the amortization burden of that machine. This can be done by creating a separate cost center for that machine and assigning a different overhead for that cost center.

Based on material, labor, and overhead, the finance department determines the standard cost for every product and measures standard cost versus actual during the year.

When it comes to make or buy decisions or large changes to the business (like expansion plans, new product opportunities, or large changes in volumes), a different approach is necessary. For make or buy decisions, justifying capital expenditures, or calculating cost savings, only variable overhead should be used and it is up to the finance department to calculate what overhead is variable and what is fixed. Fixed overhead does not change when you outsource a part, when you buy a new piece of equipment, or when you save the cost by changing methods.

The finance department should play a key role in providing financial analysis for proposals to invest in ventures or to evaluate alternative business strategies. Sometimes this requires creating a new proposed budget to see how the new plan affects the overhead, the profitability, and the cash flow.

What overhead is fixed and what is variable? From a purist standpoint, all overhead is variable. That may be true in the long run, but that does not make it right for decision-making.

Following are three examples of costing using variable overhead:

1. A company manufactures electronic systems and has a captive transformer department making most of their transformers in house. A new transformer design is up for consideration whether to make it or buy it outside. Following are two approaches of costing:

Transformer cost using standard costing:

Material = $2.50
Material burden = $1.25 (50%)
Labor = $3.00
Labor overhead = $4.50 (standard overhead is 150% of labor)

Total Cost = $11.25

Transformer cost using variable overhead:

Material = $2.50
Material burden = $1.25 (50%)
Labor = $3.00
Labor overhead = $2.40 (variable overhead = 80% of labor)

Total Cost = $9.15

Lowest outside quotes come in at $10.50. The right decision is to make the transformer in house because only variable overhead should be used for this calculation.

2. A company has a very expensive numerically controlled mill in a machine shop. It has to quote a job where all the work is done on that mill and it needs to put in a second shift for the mill. The rest of the shop works only one shift. Second shift on the mill requires an additional highly paid maintenance person; otherwise, the machine cannot work two shifts.

If the mill is not counted as a separate cost center (with its own overhead), then standard costing for the machine shop would result in the following:

Material = $20
Labor = $90 (6 hours @$15)
Overhead = $135 (150% of labor)

Total = $245

Taking into consideration the amortization of the mill and the extra maintenance overhead, plus the extra premium for second shift, the costs are quite different:

Material = $20

Labor = $112.50 (6 hours plus premium pay)

Overhead = $281.25 (250% includes amortization plus extra maintenance)

Total = $413.75

For the quote, the right cost to use is $413.75. Note that this is the cost, not the price. When calculating price, the marketing department should use the cost and add the profit.

3. A factory produces kitchen furniture. It has automated the production of its most popular line of products, but also needs to provide short runs that are not automated. The company uses a costing system that puts all burdens on labor. Due to automation of its high runners, direct labor in the factory is drastically reduced and thus indirect labor costs skyrocket. Before automation, the overhead cost was 150% of labor, but after automation it becomes 300% of labor. If the factory would use 300% overhead for all of the products, its small run custom furniture line would be penalized unfairly.

The solution is to divide the burden proportionally between labor and material. Even though the automated line uses very little labor, it would still bear a good portion of the overhead cost. Thus, the custom short run products would not be penalized because the standard lines have been automated. If all the overhead would have been left on labor, the custom products would have been priced out of the market and, by alienating customers, the standard product line sales could also be affected.

The previous examples show how important it is for financial analysts to understand the importance of correctly allocating overhead and to distinguish between fixed and variable overhead.

The preferred method of manufacturing is the continuous flow production. That means that there are no work orders and the factory works to a schedule. Once a job is opened, there are no intermediate steps until it is completed. This makes the job of financial controls easier. The alternative is job shop accounting, where for each job there is a work order that is tracked during work-in-process and there is a next step of closing the work order when the product goes to the final step (either shipping or to stores). In a job shop environment, the work order accounting cannot be avoided, but it puts a bigger burden on cost accounting.

Naturally, the decision of what method of cost accounting to use is not dependent on how easy or difficult it is to use the system, but rather the suitability of the manufacturing process. Sometimes when it is a close call, the continuous flow production should be preferred by all functions including manufacturing, materials, and accounting.

18

Human Resources and Training

Human resources must administer the wage and salary process and issue human resource policies, but they have to take care of many other functions as well. When a company is growing, new employees are hired. They come from different fields and different company cultures and they cannot be expected to know how the company operates. Young people, whether college graduates or not, may have no idea of what is expected of them, and may not have ever received training in some of the disciplines emphasized by company culture.

It is too much to expect that learning these skills can be done "on the job." It is in the company's interest to make sure that people live up to their potential and that cannot be done without some help. It is the task of human resources to provide the necessary training.

In addition, in the case of newly hired employees, there should be a "big brother" or "big sister" policy by human resources to ensure that the new employee gets smoothly settled into the company.

Human resources should have a training program that schedules various types of training during the year. Some of this has been covered under departmental responsibilities, but it bears repeating. Supervisors should get supervisory training, engineers should get training in design for manufacturability, sales people should get sales training, etc. General training classes like time management are also very useful. Human resources should have a training budget, and within this budget choose the most valuable training for the company's needs. It should keep track of who already had what training class, and it should design the frequency of classes and whether it can be held by in-house trainers, or by outside consultants.

For instance, design for manufacturability may be a class from outside consultants and may only be repeated once every three years as a refresher and to indoctrinate engineers hired during the past three years. If there are budget restrictions, a manufacturing engineer from inside the company could be tasked to research existing literature on this subject and conduct the class.

The most important training that should be given by human resources is on company culture. As we stated many times in this book, people from different walks of life and skill sets are thrown together into an organization and they need to work together as a team toward a common objective within company guidelines of behavior. These guidelines have to be explained and refreshed in their minds. This book has a lot of content about culture in a manufacturing organization, which should be summarized in a brochure by human resources and given to each new employee in addition to being the basis of his/her training class. It should include communications, conflict resolution, matrix management principles, continuous improvement programs, and personnel policies among others.

Human resources should assist in career planning. An organization's value is in its employees. Most people want to advance their skill levels. In order to retain the most valuable employees and have them live up to their potential, the company should try to help them with their career planning. Human resources should gather the information through personal reviews to understand what kind of career planning help is needed for each employee and work with their supervisors to provide that opportunity if possible. Every employee should be assisted in their career planning. For an hourly employee, it may only be cross-training, for a salaried employee it should be job related. I don't condone the company paying for classes like "Financial Training For Non-Financial People," which do very little to further company or employee personal goals.

Every training program should require for the employee to put forward some effort because it is in his/her own interest as well as the company's. If a training program is run purely on company time, it is less effective than a program where the employee has some homework or needs to do some advance preparation. For salaried employees, some of the training should not be on company time because they personally benefit from the training as well as the company. The most successful training or learning experiences require input and participation, rather than only listening. The more effort the employee puts forward, the more likely he/she will be motivated to make the most of the training.

Part of personnel philosophy of company training and career development should be whether to work on an employee's strengths or weaknesses. This question also arises during personnel reviews and supervisory training. I advocate working on strengths and just making sure that the weaknesses are shored up elsewhere. This goes along with the parallel path policies. Some outstanding employees will never make good managers. There is no use trying to shore up this weakness with management training. It is much better to use the technical skills of that individual, give him/her only technical training, and give him/her a parallel path. A good salesperson who writes reports badly will hardly benefit from a writing class, but would be much better off learning sales techniques or learning more about the product.

Building on strength gives immediate and valuable advantages, while shoring up weaknesses will be only marginally effective.

Departmental barriers often prevent people being utilized to their best advantage. Budgets sometimes prevent department managers from taking on training interns. In order to encourage training, Human Resources may allocate some of the training budget to department managers to take on training projects themselves.

In summary, the Human Resource department should have policies as follows:

- An annual schedule of training programs
- Policies for career planning for employees
- Career guidance
- Fast track for managerial talents
- Parallel path for outstanding employees without managerial talent
- Create an environment where line managers are required to train employees
- Encourage cross-training
- Recognition and incentives for employees completing training

As mentioned previously, to economize training programs, Human Resources could develop internal trainers who research the literature on their subject or go to outside classes to learn more about it, and then conduct the training classes inside the company.

19

Checklist

The following checklist can be used to determine how well a business is run, measured against the criteria in this book. The checklist is not in any order of priorities.

- Are your Board of Directors and the chair of the board independent of your CEO and key officers of the company?
- Does the board employ an independent compensation committee to determine compensation for the officers of the company?
- Is the compensation for officers set to reflect long-term goals, rather than stock price movements?
- Do you have a strategic plan for 3 to 5 years that is updated annually?
- Do you have an annual business plan?
- Does your business plan address your strengths and weaknesses versus your competition?
- Does management believe in using attainable goals?
- Does management ensure that promises on new product releases can be met?
- Does management insist that marketing promises on deliveries have been agreed to by manufacturing?
- Did your last year's business plan achieve the objectives?
- Do you have a flat organization with most decisions requiring only two signatures?
- What percentages of your hourly employees are measured against a standard?
- Do you give feedback on performance to hourly employees?
- Do you have a program for salaried employees to measure their performance against set objectives? Do you have an MBO program?

- Do you have a training program that teaches communications skills within the corporation?
- Do you have a meeting matrix schedule? Do you insist that chairs of meetings should issue an agenda and limit participants to meetings? Are you satisfied that your organization practices communications and meetings only on a need-to-know basis?
- When people come to meetings on projects, task forces, or others, do they read the minutes beforehand and come prepared to answer the questions or actions assigned to them?
- Are dates agreed upon at meetings meaningful and met most of the time, or are they just ignored and reset every time the due date is not met?
- Is marketing focused on products using product line managers, is engineering focused on design centers of excellence, is manufacturing focused on process centers of excellence, and is sales focused on regional centers?
- If your corporation is large and multinational, do you break it up into small business units by market, and does each business unit have its own profit and loss center?
- Is your corporate staff "lean and mean," meaning that they only concern themselves with auditing every business unit's performance, rather than trying to unify or standardize things within the corporation?
- Do you review your manufacturing strategy annually to find out whether you are still competitive?
- Does your manufacturing strategy include "out of the box" thinking and reviewing the latest technologies that could transform your production by using robotics or mechanization?
- Do you evaluate make or buy decisions with accounting help and use only variable overhead in this evaluation?
- Do you have a permanent cost-reduction task force that meets at least once a month with attendance by department heads from various disciplines?
- Do you plan ahead to prevent future constraints and bottlenecks within your operations? Do you have any current bottlenecks?
- Is your production organized by clusters for continuous process flow wherever possible?
- Do you have a self-administered statistical process control and output measurement charts run by your operators on the production floor?

- Is your quality function independent of production management?
- Do you have a quality education program for all employees?
- Do you measure key quality indicators and provide feedback to production?
- Are your quality personnel performing inspections or are they purely auditors?
- Do you have formal corrective action notices and procedures for quality problems for both in-house and vendors?
- Do you conduct wage and salary surveys in both local regional and national geographic areas where needed?
- Are you paying above-average wages and salaries as it applies locally or regionally?
- Do you use only basic and generic job descriptions to create grades?
- Do you use and publish a job matrix for comparison of equivalent worth?
- Do you give performance reviews and salary increases at the same time annually?
- Is your performance review format relatively painless and simple?
- Do you have a budgeted reward system for outstanding hourly and salaried employees?
- Does your engineering department have enough staff and budget not only to design new products, but also to support the cost reduction effort and sustaining engineering?
- Do you have separate new design and sustaining engineering functions?
- Do you have a training program for engineers to design for manufacturability?
- Are new designs fully tested and have pilot runs before they are released to production?
- Do design engineers spend time on the production floor? Do they play a key role in pilot running their designs in the factory?
- Does marketing have the final word on product design, pricing, and product strategy?
- Do salespersons have inflexible pricing policies that can only be changed by marketing?
- Is your monthly forecasting done according to a procedure that mandates it to be given enough consideration and to be as accurate as possible by product volume, rather than by dollars?
- Do you have an annual program with scheduled training classes that covers most employees?

- Do you have a training brochure and training program for company culture that covers most cultural and communications issues in this book?
- Do you have personnel policies for career advancement and career counseling?

Index

A

absenteeism, 63–64
activities list, 73–74
age limits, 2
airplane manufacturer, 70
Amazon company, 28–29
anniversary reviews, 84
annual reviews, 84–85
artificial turf example, 45
assembly lines, 63
AT&T, 16–17
attainable goals
    business goals, 19
    corporate culture, 30
    performance standards, 24, 98
    setting, 5
audit committee, 1
auditing, quality, 70, 72
authority with responsibility, 47, 69
awards
    issues with, 41
    outstanding performance, 103–104
    outstanding performers, 87, 101

B

backlog of inquiries, 27–28
"banana curves," 20
bell curves, 90
Bell Labs, 17
benchmarking
    geography, 81–82
    overhead, 75
benefits
    overview, 103–105
    vs. competitive wages, 53–54
"big brother/sister" policy, 121
Board of Directors
    business plan, 18
    CEO serves, 11
    role, 1–5

bonuses, 103–104, *see also* Rewards
bottlenecks, 52–53
"bottom-up" communication, 41
"bucks stops here," 11
budgeting
    annual raises, 85–86, 91
    business plans, 20
    capital expenditures, 51
    department manager training, 123
    motivators, 103
    responsiveness, 30
    zero-based, 75–76
bullpen office layout, 59–60
business plans
    annual raises budget, 85–86
    CEO role, 7–8
    correction or revision, 20
    creating, 18–20
    importance, right strategy, 16–18
    marketing perspective, 109–110
    overview, 13
    strategic plan, 2, 13–16

C

calendar, quality control, 73–74
capacity planning, 52
capital expenditures, 51–52
career planning, 122
caring, watching and, 66, 96–97
centers of excellence, 46
CEO role
    demotion/firing skills, 9–10
    most important person, 8
    overview, 7–8
    promotion skills, 9
    surviving as, 10–11
    ten commandments, 12
chain of command principles, 25–27
Chairman of the Board, 2
champion for quality, 66

change requests, responsiveness, 28
checklist of criteria, 125–128
circuit board example, 53–54
classifications, jobs, 82–83
clichés, 82
close supervision
    corporate culture, 24–25
    quality, 64
cluster work centers, 61–62
commission system, 105, 111
communications
    CEO selection, 4
    overview, 39–41
    upward and downward, 11
compensation, *see also* Wage/salary administration
    CEOs, 3
    performance standards, 24
    problems with, 53–54
compensation committee, 1
competition
    for promotion, 9
    rewards, 104
competitors
    benchmarking factors, 14
    vs. strengths and weaknesses, 15–16
completed staff work, 40–41
consensus, 7
continuous flow production, 119–120
control, span of, 32
"control limits," 78–79
controlling overhead, 75–79, *see also* Financial controls
corporate culture
    attainable goals, 30
    chain of command principles, 25–27
    close supervision, 24–25
    delivering on promises, 22–23
    discipline in workplace, 23–24
    flat organization, 32–33, 35–36
    flexibility, 33
    lack of levels, 33, 36
    mission statement, 31
    multiple levels, 33, 34
    overview, 21–22
    responsiveness, 27–30
    span of control, 32
    summary, 37
    training in, 122

corporate staff, 47–48
corrective actions, 71, 72
cosmetic appearance defects, 71
cost accounting, 116–120
cost centers *vs.* product category, 116
cost reduction, 54–56, *see also* Financial controls
creating business plans, 18–20
cubicle office layout, 60
culture, *see* Corporate culture

D

delivery
    competitors, 14
    on promises, 22–23
demotion/firing, 9–10
design for manufacturability, 107–108, 121
design stability, 77
Detroit's quality problems, 63–64
dignity, 10
direct labor
    financial controls, 116
    overhead, 75–76
director, *see* Board of Directors; CEO role
discipline in workplace, 23–24
"dis-satisfiers," 82
diversification, 15

E

educational needs/ambitions, 92, *see also* Training
80/20 rule, 79
electronics industry example, 71
elephant gift, 76
e-mail etiquette, 39
engineering
    change requests, 28
    operator charts, 79
    organization of, 45
    role, 22, 107–108
environment, 97
equipment
    capital expenditures, 51–52
    duplicating, 62
    modern and competitive, 51
    ribbons on, 96

evaluation of performance, 84–87
excuses, poor responsiveness, 30
exit strategy, 3, *see also* Succession plans

F

failure, 9
feedback, 71, 72, 92
financial controls, *see also* Controlling overhead
    cost reduction, 54–56
    overview, 115–120
firing/demotion, 9–10
flat organization
    corporate culture, 32–33, 35–36
    span of control, 32
flexibility, 33
"focused factories," 50
forecasting, 113–114
free stock, production floor, 56
fringe benefits, *see* Benefits

G

Geely Holdings, 63
General Electric factory experiment, 96–97
geography, 81–82
global environment
    competitors, 14
    divisions in organizations, 45–46
goals, *see* Attainable goals
"golden parachute," 3
governance committee, 1
grades (levels), 82–83, 100–101
grooming successors, 5

H

Hawthorne experiment, 66, 96–97
headset example, 16–17
"hockey stick" annual plan, 20
hourly workers
    motivation, 95, 99
    performance reviews, 90
    reward system, 104
    turnover, 81
human resources, 121–123

I

incentives, 103–105
incomplete objectives, 101
indirect labor, 75–76
information overload, 40
inquiries backlog, 27–28
inspections, incoming, 70, 72
International Organization for Standardization, 67
Internet use, 39
ISO 2000/9000, 67
Israel (state), 76
IT requests, responsiveness, 28

J

Japan, *see* Quality circles
job descriptions and classifications, 82–83
"job shop" environment, 62
just-in-time systems, 56, 59

K

Kanban systems, 56, 77, 78
"kingdoms," 33
kitchen furniture example, 119

L

labor contracts, 17
lack of levels, 33, 36
laissez-faire management, 23–24
large corporations, organization of, 43–48
lateral moves, 10
lawsuit protection, 92
Lean Manufacturing, 65
levels in organization, 32–36
lights-out factory, 75
lightweight headset example, 16–17
listening sessions, 41, 90, 100
"little kingdoms," 33
location
    centers of excellence, 46
    competitors, 14
    outsourcing/make or buy, 50–51
    strategic decision, 17
lonely life, 11

M

machine shop example, 118–119
make or buy
    financial controls, 117
    manufacturing strategies, 49–51
management
    fads, 21
    laissez-faire, 23–24
    levels in organization, 32–36
management by exception, 79
management by objectives (MBO)
    cost reduction, 54
    reward system, 104
    salaried employees, 90, 99–100
managers, CEO difference, 7
manufacturability, design for, 107–108, 121
manufacturing
    constraints manufacturing strategies, 52–54
    organization of, 45
    overview, 61–67
manufacturing strategies
    capital expenditures, 51–52
    cost reduction, 54–56
    make or buy, 49–51
    manufacturing constraints, 52–54
    materials systems, 56–59
    office layouts, 59–60
    outsourcing, 49–51
    overview, 49
    supply chain management, 56–59
market-driven organizations, 44–45
marketing and sales, *see also* Sales
    CEO selection, 4
    competitors, 14
    overview, 109–114
    role in success, 22
materials systems, 56–59
matrix, job grades, 83
"maverick" spirit, 24
meetings
    attendance, 2
    forecasting, 113–114
    schedule, 40
"me too" approach, 16
Mexico
    labor contracts, 17
    promotion selection, 64

Microsoft company, 29
milestones, 19
min-max system, 57
missionary sales, 112
mission statement, 31
Monday absenteeism, 63–64
motivation, 95–101
Motorola, 65
multiple levels, 33, 34

N

90/10 rule, 79
Northern Telecom (Nortel), 17–18, 26, 53
numerically controlled mill example, 118–119

O

office layouts, 59–60
operator charts, 79
organization of large corporations, 43–48
outsourcing and offshore strategy, 49–51
overhead, control of
    financial controls, 116
    outsourcing/make or buy, 50
    overview, 75–79
Owens, Robert, 96

P

Pareto Rule, 79
participating principles, 7
partitions, office layout, 60
parts, product redesign, 77–78
pay grades, *see* Compensation
perfection, 70, 71
performance reviews, 87, 89–93
permissiveness, 24
personal recognition, 97
pictures, unacceptable quality, 71
piecework, 97, 104
Plantronics, 16–17
plastic cover example, 71
plumbing manufacturer, 70
Prime Minister of Israel, 76
process-oriented work centers, 61–62
production management and managers, 70–71, 110–111

production workers, 62–63
productivity
    office layout, 60
    overview, 95–101
    reports, 78
product redesign, 77–78
products
    categories *vs.* cost centers, 116
    competitors, 14
    introducing new, 108
    life cycle, 110
    manufacturing the right, 18
    proprietary, 50
profit centers, 32, 115
promises, responsiveness, 28
promotion
    CEO role, 9
    selection, 111–112
proprietary products, 50
protection of turf, 24

Q

quality circles, 65–66
quality control
    absenteeism, 63–64
    calendar, 73–74
    extra effort, cost, 70–71
    overview, 69–74
    pictures of unacceptable, 71
    plan, 70

R

raises, *see* Annual reviews; Budgeting
random sampling, 72
razorblade's edge example, 71
redesign of products, 77–78
repair operation example, 27–28
reports
    costs, 78
    performance objectives, 98–99
    product category *vs.* cost centers, 116
resource allocation, 20
responsibility with authority, 47, 69
responsiveness, 27–30
rewards
    issues with, 41
    outstanding performers, 87, 101, 103–104

ribbons example, 96
right strategy importance, 16–18
roles
    Board of Directors, 1–5
    CEO, 7–12
    engineering, 107–108
Rolm company, 54
rooms within rooms, 76
rural manufacturing company example, 59

S

salaried workers
    motivation, 99–100
    performance reviews, 90
salaries, *see* Wage/salary administration
sales, 45, *see also* Marketing and sales
scientists' reports, 98–99
semiconductor company example, 44
sense of achievement/belonging, 97
shifts example, 118–119
Siam (ruler of), 76
sideways moves, 10
Six Sigma, 65
size, competitors, 14
slogans, 69
small business units (SBUs), 43, 46
social experiments, 62–63
social factors, 97
social functions, 76–77
span of control
    CEO role, 8
    corporate culture, 32
spare parts, 53
stability in design, 77
standards, 67, 98
stock options, 105
strategic plans, 13–16
strategy, importance of right, 16–18
strengths and weaknesses
    CEOs, 4
    employee's, 122–123
    vs. competition, 15–16
    working on, shoring up, 112
succession plans, 5, *see also* Exit strategy
suggestion plan programs, 41
"Superman" CEO, 4

supervision
    close, corporate culture, 24–25
    number of employees, 64
supply chain, 14
supply chain management, 56–59
switching division (Nortel), 26–27
"SWOT" system, 57

T

teamwork
    environment, 10
    sales forecasting effect, 114
    training in, 24
telephone use, 40
ten commandments, 12
termination, 92
term limits, 2
Tesla automobile example, 23
360 degrees, 91
tie clip example, 55
timetables
    business plan, 19
    quality calendar, 73–74
time to market, 53
Titanic example, 21
titles, job classifications, 82–83
Total Quality Management (TQM), 65
Toyota, 65
TQM, *see* Total Quality Management (TQM)
track record, CEOs, 3
trade barriers, 14
training
    communications, 39
    company culture, 122
    department managers, 123
    engineers, 107
    meetings, 40
    overview, 121–123
    performance reviews, 92
    quality work, 70–71
    responsiveness, 29
    teamwork, 26
transactions, 77–78
transformer design example, 117–118
Truman (President), 11
turf protection, 24
Twain, Mark, 113

U

unacceptable quality, pictures, 71
undermining, 9–10
"understudy" successors, 5

V

value analysis, 55
variances, 78–79
vendors, 57–58
visual aids, 71
Volvo example, 62–63

W

wage/salary administration, *see also* Compensation
    annual budget, 85–86
    competitive *vs.* benefits, 53–54
    overview, 75–79
    vs. fringe benefits, 103
watches example, 44
watching and caring, 66, 96–97
work centers, 61–62
working conditions, 97
writing up employees, 10

Z

zero-based budgeting, 75–76
"Zero Quality Program," 65

About the Author

Frederick Parker was born in Hungary and escaped from the communist rule in 1949.

He immigrated to Australia and received his degree in Industrial Engineering at the University of New South Wales.

After coming to America he specialized in manufacturing management and turn-around situations for several hi-tech companies, including corporate strategies for international companies.

Parker was general manager of a large international telecommunication company in Silicon Valley, in charge of 1,200 employees, shipping $500 million of hi-tech equipment from a 300,000-square-foot factory. He achieved 55% gross margins. When he was promoted to corporate vice president of operations, he was asked to research and formulate the manufacturing strategy for this international corporation. At that time the company had factories all over the world. In that strategic role Parker studied the latest trends in world-class manufacturing before implementing his plans for the corporation.

Parker also built and managed factories in Mexico and later as Vice President of Operations took part of taking a company public. The year after that public offering, the company's stock was the largest percentage gainer on the NASDAQ stock exchange.

Before retiring and writing this book, Parker consulted for several companies that relied on his expertise of New Product Introduction and Cost Reduction.

Strategy +Teamwork is based on his 40 years of experience in manufacturing management and his strong desire to pass on the wisdom he gained in confronting difficult problems in the competitive global environment.

Parker is consulting part-time and lives in San Diego with his wife and has two children and three grandchildren.

He also wrote a book on bridge, *Win at Duplicate Bridge*, and he enjoys playing bridge, golf, chess, and writing articles.